Ponnadurai Ramasami (Ed.)
Computational Chemistry Methods

Also of interest

Computational Sciences
Ramasami (Ed.), 2017
ISBN 978-3-11-046536-5, e-ISBN 978-3-11-046721-5

Density Functional Theory
Advances in Applications
Ponnadurai Ramasami (Ed.), 2018
ISBN 978-3-11-056675-8, e-ISBN 978-3-11-056819-6

Chemoinformatics of Natural Products
Fundamental Concepts
Fidele Ntie-Kang (Ed.), 2020
ISBN 978-3-11-057933-8, e-ISBN 978-3-11-057935-2

Chemoinformatics of Natural Products
Advanced Concepts and Applications
Fidele Ntie-Kang (Ed.), 2020
ISBN 978-3-11-066888-9, e-ISBN 978-3-11-066889-6

Physical Sciences Reviews.
e-ISSN 2365-659X

Computational Chemistry Methods

Applications

Edited by Ponnadurai Ramasami

DE GRUYTER

Editor
Prof. Dr. Ponnadurai Ramasami
Computational Chemistry Group, Department of
Chemistry, Faculty of Science, University of Mauritius,
Réduit 80837, Mauritius
Department of Chemistry, College of Science,
Engineering and Technology, University of South Africa,
P.O. Box 392, Pretoria 0003, South Africa
E-mail address: p.ramasami@uom.ac.mu

ISBN 978-3-11-062906-4
e-ISBN (PDF) 978-3-11-063162-3
e-ISBN (EPUB) 978-3-11-062992-7

Library of Congress Control Number: 2020938612

Bibliographic information published by the Deutsche Nationalbibliothek
The Deutsche Nationalbibliothek lists this publication in the Deutsche Nationalbibliografie; detailed bibliographic data are available on the Internet at http://dnb.dnb.de.

Cover image: noLimit46 / iStock / Getty Images Plus
Typesetting: Integra Software Services Pvt. Ltd.
Printing and binding: CPI books GmbH, Leck

www.degruyter.com

Preface

A virtual conference on computational science (VCCS-2018) was organized online from 1st to 31st August 2018. This was the sixth virtual conference which was started in 2013. The month of August was chosen to commemorate the birth anniversary of Erwin Schrödinger, the father of quantum mechanics, on 12th August.

There were 30 presentations for the virtual conference with 100 participants from 20 countries. A secured platform was used for virtual interactions of the participants. After the virtual conference, there was a call for full papers to be considered for publication in the conference proceedings. Manuscripts were received and they were processed and reviewed as per the policy of De Gruyter.

This book is a collection of the eight accepted manuscripts based on the use of computational chemistry methods. These manuscripts cover a range of topics from fundamental to applied science. Choong et al investigated the conformations and interactions between *R*- and *S*-methandone in wild type CYP2B6, 2D6 and 3A4. Gümüş et al studied the aromaticity of mono, di, tri and tetraazaphenantherene derivatives. Renita and Sivasubramanian reviewed the application of computational chemistry for adsorption studies on metal organic frameworks for carbon capture. Kuznetsov performed a DFT study on phthalocyanines core-modified by P and S and their complexes with C60. Kharkar et al presented models for the computational predication of toxicity of small organic molecules. Kakkar et al reviewed a combined approach of homology modeling, molecular dynamics and docking towards computer-aided drug discovery. Ramasami et al investigated the structural and spectroscopic parameters of 2,4-dichloro-N-phenethylbenzenesulfonamide and 2,4-dimorpholino-4-yl-6-(4-nitrophenoxy)-[1,3,5]-triazine in combined experimental and DFT studies.

I hope that these chapters will add to literature and they will be useful references.

To conclude, VCCS-2018 was a successful event and I would like to thank all those who have contributed. I would also like to thank the Organising and International Advisory committee members, the participants and the reviewers.

Prof. Ponnadurai Ramasami

https://doi.org/10.1515/9783110631623-202

Contents

List of contributing authors

Surendra Agrawal
Shobhaben Patapbhai Patel School of Pharmacy and Technology Management,
SVKM's NMIMS (Deemed to be University)
Vile Parle (W)
Mumbai 400 056 India
Surendra.Agrawal@nmims.edu

Varun Chahal
Computational Chemistry Laboratory
Department of Chemistry
University of Delhi Faculty of Science
Delhi 110007, India
chahal.varun94@gmail.com

Yee Siew Choong
Institute for Research in Molecular Medicine
Universiti Sains Malaysia
Penang, Malaysia
yeesiew@usm.my

Chaitannya W. Ghanavatkar
Department of Dyestuff Technology,
Institute of Chemical Technology
Matunga, Mumbai 400 019
India

Ayşegül Gümüş
Faculty of Science
Department of Chemistry
Van Yuzuncu Yil University
Van 65080
Turkey
gumusa@gmail.com

Selçuk Gümüş
Faculty of Science
Department of Chemistry
Van Yuzuncu Yil University
Van 65080
Turkey
gumuss@gmail.com

Zeynep Turhan İrak
Faculty of Engineering
Department of Environmental Engineering
Igdir University
Igdır Unv. Sehıt Bulent Yurtseven Campus,
Suveren, Merkez
Igdir 76000, Turkey
zeynepsilanturhan@hotmail.com

Rita Kakkar
Computational Chemistry Laboratory
Department of Chemistry
University of Delhi Faculty of Science
Delhi 110007, India
rkakkar@chemistry.du.ac.in

Nik Nur Syazana Bt Nik Mohamed Kamal
Institute for Research in Molecular Medicine
Universiti Sains Malaysia
Penang, Malaysia

Necdet Karakoyun
Faculty of Science
Department of Chemistry
Van Yuzuncu Yil University
Van 65080
Turkey
necdetkarakoyun@yyu.edu.tr

B. Karthikeyan
Department of Chemistry
Annamalai University
Annamalai Nagar
Tamilnadu, India
bkarthi_au@yahoo.com

R. Kavipriya
SRM Institute of Science and Technology
Ramapuram campus, Department of Chemistry
600089 Chennai, India
mkrkavipriya@yahoo.co.in

Helen P. Kavitha
SRM Institute of Science and Technology
Ramapuram campus, Department of
Chemistry
600089 Chennai, India
helenkavithap@yahoo.co

Prashant S. Kharkar
Shobhaben Patapbhai Patel School of
Pharmacy and Technology Management,
SVKM's NMIMS (Deemed to be University)
Vile Parle (W)
Mumbai 400 056 India
Prashant.Kharkar@nmims.edu

Aleksey E. Kuznetsov
Departamento de Química
Universidad Técnica Federico Santa María
Av. Santa María, Vitacura
Santiago 6400
Chile
aleksey73kuznets@gmail.com

Theam Soon Lim
Institute for Research in Molecular Medicine
Universiti Sains Malaysia
Penang, Malaysia

Janvhi Machhar
Shobhaben Patapbhai Patel School of
Pharmacy and Technology Management,
SVKM's NMIMS (Deemed to be University)
Vile Parle (W)
Mumbai 400 056 India
janvhimachhar97@gmail.com

Virendra R. Mishra
Department of Dyestuff Technology,
Institute of Chemical Technology
Matunga, Mumbai 400 019
India

Ansh Mittal
Shobhaben Patapbhai Patel School of
Pharmacy and Technology Management,
SVKM's NMIMS (Deemed to be University)
Vile Parle (W)
Mumbai 400 056 India
anshmittal@live.com

Sonam Nirwan
Computational Chemistry Laboratory
Department of Chemistry
University of Delhi Faculty of Science
Delhi 110007, India
sonamnirwan09@gmail.com

Erhan Öztürk
Faculty of Science
Department of Chemistry
Van Yuzuncu Yil University
Van 65080
Turkey
kimyagerhan@hotmail.com

Anil M. Pethe
Shobhaben Patapbhai Patel School of
Pharmacy and Technology Management,
SVKM's NMIMS (Deemed to be University)
Vile Parle (W)
Mumbai 400 056 India
Anil.Pethe@nmims.edu

Ponnadurai Ramasami
University of Mauritius, Department of
Chemistry, Computational Chemistry Group,
Faculty of Science, 80832 Réduit, Mauritius,
ramchemi@intnet.mu

A. Annam Renita
Department of Chemical Engineering
Sathyabama Institute of Science and
Technology
Chennai 600119, India
annamrenita.chem@sathyabama.ac.in

Lydia Rhyman
University of Mauritius, Department of
Chemistry, Computational Chemistry Group,
Faculty of Science, 80832 Réduit, Mauritius,
lyd.rhyman@gmail.com

Nagaiyan Sekar
Department of Dyestuff Technology,
Institute of Chemical Technology
Matunga, Mumbai 400 019
India
nethi.sekar@gmail.com

Vandana Kumari Shukla
Department of Dyestuff Technology,
Institute of Chemical Technology
Matunga, Mumbai 400 019
India

V. Sivasubramanian
Department of Chemical Engineering
National Institute of Technology
Calicut 673601
Kozhikode, India

Jasmine P. Vennila
Department of Physics
Panimalar Institute of Technology, Chennai
Tamilnadu, India
jasprabhu1@yahoo.co.in

Erhan Öztürk, Zeynep Turhan İrak, Necdet Karakoyun, Ayşegül Gümüş and Selçuk Gümüş

1 Investigation of the aromaticity of mono, di, tri and tetraazaphenanthrene derivatives

Abstract: In this chapter mono, di, tri and tetraaza substituted phenanthrene derivatives have been investigated computationally with B3LYP/6-31 + G(d,p) level of theory. Substitution of carbon atom of the main structure with nitrogen obviously disturbs the aromaticity, indeed it decreases it. Thus, the idea of regaining of the aromaticity back by using electron withdrawing groups came across. As a result of the computational calculations, energetically most unfavored structures have been found to be those where aza substitutiona are vicinal. Secondly, the aromaticities of the present species depend on the position of the centric substituent. In addition, the effect position of the side substituent has been considered. The system becomes more aromatic (possess greater negative NICS values or smaller HOMA value) when the electron withdrawing atoms or groups are adjacent to the centrically substituted heteroatoms.

Keywords: aromaticity, NICS, HOMA, phenanthrene, azaphenanthrene

1.1 Introduction

The polycyclic aromatic hydrocarbon formed by fusing three benzene rings is called Phenanthrene (**Ph**) which takes its name from two well-known chemical compounds; 'phenyl' and 'anthracene'. is It has a role as an environmental contaminant and a mouse metabolite. It is an ortho-fused polycyclic arene, an ortho-fused tricyclic hydrocarbon and a member of phenanthrenes (Figure 1.1). The steroids possess phenanthrene in their framework. It is a harmfull material in its pure form, which is spread out by cigarette smoke. Its irritant and photosensitising skin to light properties are well known. Phenanthrene emerges as a white powder chemical emitting blue fluorescence. Although is insoluble in water its solubility in most organic solvents, including toluene, carbon tetrachloride, ether, chloroform, acetic acid and benzene, is quite high. Bardhan-Sengupta Phenanthrene Synthesis is the classical synthetic application to obtain **Ph** [1].

This article has previously been published in the journal Physical Sciences Reviews. Please cite as: Öztürk, E., Turhan İrak, Z., Karakoyun, N., Gümüş, A., Gümüş, S. Investigation of the aromaticity of mono, di, tri and tetraazaphenanthrene derivatives *Physical Sciences Reviews* [Online] 2020, 5 DOI: 10.1515/psr-2018-0176.

https://doi.org/10.1515/9783110631623-001

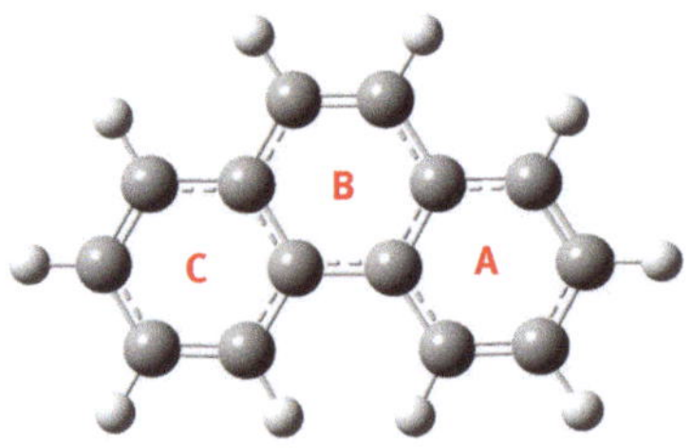

Figure 1.1: Structure of phenanthrene and naming of the rings.

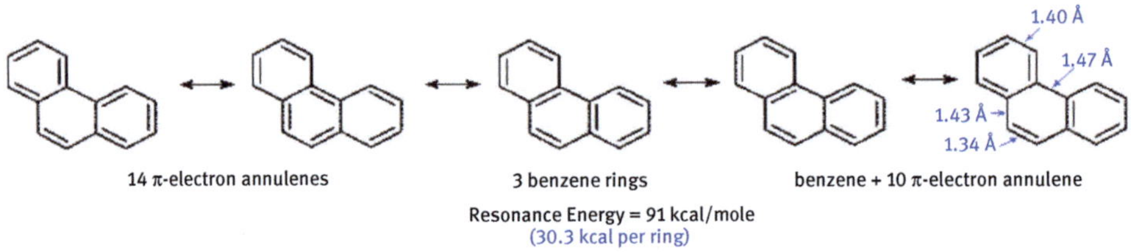

Figure 1.2: Resonance structures of phenanthrene.

Phenanthrene has a significant resonance stabilization due to its tricyclic aromatic structure, as shown in the following diagram (Figure 1.2). Three fused benzene ring moieties is one of the five contributing resonance structures, in two of the structures 10 π-electron annulene is fused to a benzene ring, and the remaining two geometries are 14 π-electron annulenes. All of the resonans structures are aromatic by the https://www2.chemistry.msu.edu/faculty/reusch/virttxtjml/react3.htm#rx9ac Hückel Rule. A careful inspection of each contributing structure reveals that the carbon-carbon bond of the B–Ring has 80 % double bond character (1.34 Å bond length), whereas the opposite bond across the circle has 80 % single bond character with 1.47 Å bond length. The bond lengths shown on the right hand structure reflected that the other carbon-carbon bonds vary in terms of bond order. As expected, the carbon–carbon bonds with more than 80 % double bond character, display double bond-like addition reactions, including simple catalytic hydrogenation. Indeed both radical and polar addition reactions are observed by all the aromatic fused ring compounds which less readily benzene undergoes.

Kalescky et al. [2] introduced a new description for the aromaticity of polycyclic compounds including phenantherene. They compared the aromatic character of the compounds in terms of bond strengths obtained from computations of vibrational frequencies. A similar argument was concluded for phenanthrene since the bond with double bond character was calculated to be much more stronger than the bond with single bond character [2].

Apart from the **Ph** itself azaphenanthrene derivatives have been considered by the researchers both experimentally and theoretically [3–7]. The most popular azaphenanthrene in the literature is 1,10-phenanthroline [7]. 1,10-Phenanthroline (**Ph_1_10**) took part very important role in the development of coordination chemistry due to being a classic chelating bidentate ligand for transition metal cations [8–10]. Considerable interest to it as adaptable starting material for organic, inorganic and supramolecular chemistry still continues. **Ph_1_10** possess rigid planar, hydrophobic, electron-poor heteroaromatic character whose nitrogens are located nicely to act together in cation coordination. Thus, its binding ability toward metal ions is determined by these structural features [9, 10].

Research on aromaticity has still been an intensively investigated area of chemistry. The Huckel rule states that; if monocyclic aromatic compounds with cyclic conjugated π-systems contain the proper number of π-electrons, they are considered to be aromatic. Although these criteria are strong enough to decide the aromaticity of neutral and charged ring systems, some additional definitions are necessary to indicate the aromaticity of more complex systems clearly (as in our case).

In general, aromaticity can be expressed by a combination of terms such as energetic, structural and magnetic criteria in cyclic delocalized systems [11–16]. Nucleus-Independent Chemical Shift (NICS) has been introduced by Schleyer in 1996, which is a simple and efficient probe for aromaticity [17]. NICS is the computationally calculated value of the negative magnetic shielding at some selected point in space of the molecule, generally, at a ring or cage center. Aromaticity is denoted by negative NICS data (−11.5 for benzene, −11.4 for naphthalene) whereas positive values of NICS denote antiaromaticity (28.8 for cyclobutadiene). In addition, nonaromatic systems possess small NICS values (−3.1 for 1,3-cyclopentadiene). NICS is proven to be a convenient indicator of aromaticity that often correlates very well with the other energetic, structural and magnetic criteria [18–21]. Overall of a polycycle can be represented by resonance energies and magnetic susceptibilities, however they do not provide information about the individual rings. Fortunately, local aromaticity of individual rings of polycyclic compounds are judged by computing NICS, which makes it be an effective probe for aromaticity.

Another index for aromaticity, which is geometry-based, is named as Harmonic oscillator measure of aromaticity (HOMA) [14]. HOMA is different from the other geometry-based indices in terms of consideration of a reference bond length or bond order. In HOMA model an idea of the optimal bond length is taken as a reference instead of the mean bond. HOMA data are between 0 and 1. A non-aromatic system gets 0 HOMA value, on the other hand, 1 is obtained for a system where full π-electron delocalization occurs (benzene). The extent of π-electron delocalization (aromatic character) of six membered heterocycles can be quantified by HOMA.

The following formula can be used to calculate HOMA:

$$\mathrm{HOMA} = 1 - \frac{\alpha}{n}\sum (R_{\mathrm{opt}} - R_i)^2 \tag{1.1}$$

$$= 1 - \frac{\left[\alpha_{CC}\sum\left[R_{(CC)opt} - R_i\right]^2 + \alpha_{CX}\sum\left[R_{CXopt} - R_i\right]^2\right]}{n} \tag{1.2}$$

In this equation, α is the normalization constant and n is the number of chemical bonds considered. The optimum bond length is represented by R_{opt} that is obtained when full delocalization of π-electrons, and R_i are the computed bond lengths. For CC bonds α is fixed to 257.7 which gives HOMA = 0 for a non-aromatic system (e. g. Kekule´ structure of benzene), and HOMA = 1 is found for an aromatic compound whose all the bonds are computed to be equal to the optimal bond length value.

For CC bonds, R_{opt} is assumed to be equal to 1.388 Å when the compound is realized as a full aromatic system. PDI, ATI and FLU indices [22] and HOMA index [23–25] have been studied in the literature to decide the relationship between substituent effect and the aromatic nature of aza analogs of naphthalenes, recently.

In this chapter, the aromaticity of mono, di, tri and tetraazaphenanthrene derivatives was computed by the application of Density Functional Theory using the B3LYP hybrid functional and 6-31G+(d,p) basis set. Introduction of an heteroatom, nitrogen in the present case, creates azaphenanthrenes. By means of centric perturbation of one, two, three and four nitrogens at different positions, mono, di, tri and tetraazaphenanthrene derivatives are structurally obtained from phenanthrene, respectively. It is obvious that the introduction of nitrogens will reduce the aromaticity of the parent phenanthrene structure due to less effective electron localization because of the disturbed ring current arising from electronegativity of the nitrogens at the perturbation site. The goal of the present research article was to study the substituent effect on the aromaticity of azaphenanthrenes by means of NICS and HOMA indices. The effect of the position of NO_2 substituent on the present systems has been theoretically studied by the application of density functional theory (DFT) calculations focusing especially on NICS data. Moreover, HOMA indices have been calculated by the formula for all the compounds to judge the aromaticities. Lastly, a comparison has been done with NICS and HOMA data to gain inside whether magnetic or geometry-based criteria better defines the aromatic characters.

1.2 Method of calculation

Semi-empirical PM3 self-consistent field molecular orbital (SCF MO) method has been applied for initial geometry optimizations for all the structures leading to energy minima [26, 27] at the restricted level [28]. Afterwards, geometry optimizations were performed within the framework of density functional theory (DFT, B3LYP) [29, 30] at the level of 6-31G(d,p) (restricted closed-shell) [27]. B3LYP is an exchange term consisting of hybrid Hartree–Fock and local spin density (LSD) functions with

Becke's gradient correlation to LSD exchange [31]. Vosko, Wilk, Nusair (VWN3) local correlation functional [32] and Lee, Yang, Parr (LYP) correlation correction functional [33] construct the correlation term of B3LYP. SCF-HF results are improved by the B3LYP method whose predictions are in qualitative agreement with experimental data [34–36].

The normal mode analysis performed for all the structures. Each structure does not yield any imaginary frequencies for the 3N-6 vibrational degrees of freedom, where N is the number of atoms in the system, which indicates that the geometry of each molecule represents a local minimum on the potential energy surface. NMR shielding data [37] were computed using the Gauge-Independent Atomic Orbital method [38] with the restricted closed shell method performing 6–31 + G(d,p) basis set over B3LYP/6-31G(d,p) optimized geometries. NICS values were calculated theoretically by obtaining NMR shielding at the ring centers, NICS(0). Gaussian 09 W package program was used for the geometry optimizations and NICS calculations of the present systems [39].

1.3 Results and discussion

Substituted phenazine cores are usually found in natural products, dyestuffs, pesticides, and antibiotics, which makes them important biologically active motifs. A majority of them are produced naturally by bacteria from diverse genera including Pseudomonas, Pelagiobacter, Vibrio and Streptomyces species [40].

Both theoretical and experimental studies on the effect of centric substitution of an heteroatom to the parent ring, and/or substitution of an heteroatom or heterogroup with the outer hydrogens of aromatic molecules have always found application in the literature. In the present article, mono, di, tri and tetraazaphenanthrene derivatives and their substituted (**NO_2**) counterparts have been investigated theoretically by the application of B3LYP/6-31 + G(d,p) level of theory in order to decide their stabilities and aromaticities.

The mono and/or dicentric perturbation on positions of ring fusions are not taken into account since non-aromatic systems are resulted in those cases. The molecules were named according to the positions of the nitrogens on the system from **Ph_1** to **Ph_1_2_3_4** for tetraazaphenanthrene (see Figure 1.3). **Ph** represent phenanthrene and the numbers indicate the position of the heteroatom substitution.

1.4 Energetics

The aforementioned method has been used to obtain the zero point corrected total electronic energies of the present compounds. The total energies (in a.u.) and corresponding relative energies (in kJ/mol) of the systems are given in Table 1.1. The

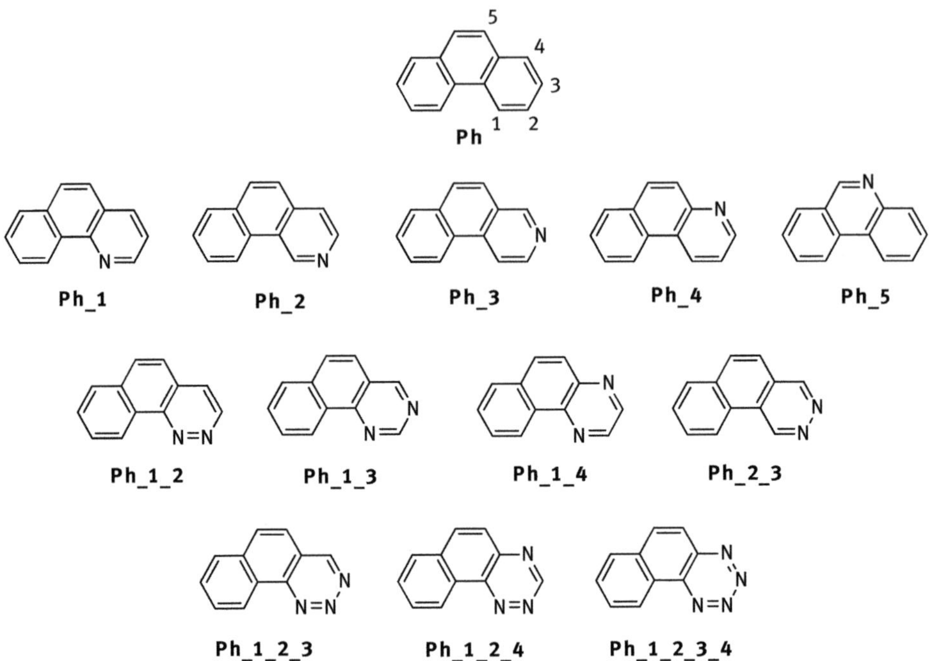

Figure 1.3: Chemical structures of mono, di, tri and tetraazaphenanthrene derivatives. The numbering in **Ph** indicates the positions of aza substitutions.

Table 1.1: Total energies (a.u.) and corresponding relative energies (kJ/mol) of the two series.

Structure	Total Energy	Relative Energy
Ph	−539.53865526	
Ph_1	−555.58020638	0
Ph_2	−555.57523782	12.55
Ph_3	−555.57589079	10.90
Ph_4	−555.57711401	7.81
Ph_5	−555.57775942	6.18
Ph_1_2	−571.58245993	95.11
Ph_1_3	−571.62011790	0
Ph_1_4	−571.61467372	13.75
Ph_1_5	−571.61356400	16.55
Ph_2_3	−571.61686438	8.22
Ph_2_4	−571.61873150	3.50
Ph_2_5	−571.61353711	16.62
Ph_3_4	−571.61347597	16.77
Ph_3_10	−571.61314591	17.61
Ph_1_2_3	−587.60516311	64.76
Ph_1_3_4	−587.63080594	0
Ph_1_2_3_4	−603.60077732	

lowest energy compounds have been found to be **Ph_1, Ph_1_3 and Ph_1_2_4** for mono, di and triazaphenanthrene derivatives, respectively.

Vicinal placement of two or three nitrogen atoms results in drastic instability **Ph_1_2** and **Ph-1-2-3** over the other derivatives, which disturbs the aromaticity of the related system much more than the other centric substitutions. On the other hand, the position of the nitrogen substitutions has very little effect on the total energy for monoaza derivatives.

1.5 Nucleus-Independent Chemical Shift

The contradictions and paradoxes between different measures of aromaticity have been described by Stanger in the literature [41].

The delocalization of Huckel number of π-electrons conjugatively in a ring accounts for the aromaticity in that ring which leads to better stability. NICS is an outcome magnetic properties of the ring under consideration. There exists a perfect delocalization of six π-electrons in benzene which is the most well-known aromatic compound. Therefore, central substitution of an heteroatom is expected to decrease the aromaticity of the ring to some extent due to the electronegativity difference between carbon and heteroatoms. Indeed, substitution of a second heteroatom even decreases the aromaticity of that ring more as in the present case. The NICS data for the parent phenanthrene, mono, di, tri and tetraaza substituted phenanthrene derivatives are given in Table 1.2.

The data show what was being expected. For **Ph** the ring in the middle computed to be less aromatic than the side rings as expected through the argument given in the introduction part. Centric substitution of nitrogen decreases the aromaticity of that ring. The aromaticity of the A ring of the monoazasubstituted rings decreased around −1.5 from the parent **Ph.** Moreover, a NICS data of −3.71 ppm have been obtained for **Ph_1_2_3_4** leading to ring of almost non-aromatic character.

If aromaticity is considered as an additive property for the compounds with fused rings, one can sum up the aromaticity of each ring to obtain a total aromaticity of the compound. Herein the total NICS data for parent **Ph** has been computed to be −27.05 ppm. After aza substitution the total aromaticity of the system decreases reaching to a 7 ppm decrease for the tetraazaphenanthrene.

However, the lost aromaticity can be gained back by the replacement of ring hydrogens by an electron withdrawing atom or group. The powerful electronegative property of the substituent makes pulling of the electrons located on the heteroatom forming the central ring back into the circle to improve the aromaticity. In our case, the effect of substitution of strongly electron withdrawing $\mathbf{NO_2}$ group on the aromaticity of the azaphenanthrenes has been investigated by computing the NICS values at the ring centers (NICS(0)).

Table 1.2: NICS (ppm) data for the present systems.

Structure	A Ring	B Ring	C Ring	Total
Ph	−10.06	−6.87	−10.12	−27.05
Ph_1	−8.55	−6.79	−10.13	−25.47
Ph_2	−8.65	−6.92	−10.13	−25.70
Ph_3	−8.54	−6.98	−10.14	−25.66
Ph_4	−8.56	−6.73	−10.13	−25.42
Ph_5	−10.10	−5.42	−10.13	−25.65
Ph_1_2	−7.16	−6.41	−9.98	−23.55
Ph_1_3	−7.04	−6.95	−10.14	−24.13
Ph_1_4	−6.97	−6.55	−10.12	−23.64
Ph_1_5	−8.51	−5.34	−10.20	−24.05
Ph_2_3	−6.27	−7.63	−10.21	−24.11
Ph_2_4	−7.12	−6.85	−10.16	−24.13
Ph_2_5	−8.75	−5.33	−10.10	−24.18
Ph_3_4	−7.12	−6.39	−10.05	−23.56
Ph_3_5	−8.43	−5.48	−10.25	−24.16
Ph_4_5	−8.78	−5.47	−10.17	−24.42
Ph_1_2_3	−5.15	−7.13	−10.20	−22.48
Ph_1_2_4	−5.49	−6.39	−10.10	−21.98
Ph_1_2_3_4	−3.71	−6.34	−10.11	−20.16

According to the results of NICS calculations of substituted mono, di, tri and tetraaza substituted phenanthrene derivatives, it is clear that the system becomes more aromatic by the substitution which confirms the explanation above. For all cases, the aromaticity of the unsubstituted aza derivatives of phenanthrene has been increased by the substitution of a very strong electron withdrawing nitro group. For example, the A ring of the unsubstituted triazaphenanthrene compounds **Ph_1_2_3** and **Ph_1_2_4** have NICS values of −5.15 and −5.49 ppm, respectively. The lost aromaticities of these system has been enhanced to −7.13 and −9.30 ppm upon substitution of **NO_2**, respectively. NICS data for different positions of the substituents, when substitution is closer the aza points of the rings, the aromaticity of the increases. Inversely, NICS of the systems becomes smaller (in absolute value) while the electronegative group situated farther (Figure 1.4).

1.6 Harmonic oscillator measure of aromaticity

The calculated HOMA data for the azaphenanthrene derivatives are reported in Table 1.3. Careful inspection of the table shows that; geometry index HOMA data for the aza substituted rings present similar effect of substituents on the aromaticity as in the case of magnetic property index, NICS. The argument explained above for NICS can hold for the present aromaticity definition. The HOMA data decreased from 0.898 to 0.538 going from **Ph** to **Ph_1_2_3_4**. The reducing effect of nitrogen substitution on the aromaticity could be clearly observed by HOMA calculations as well.

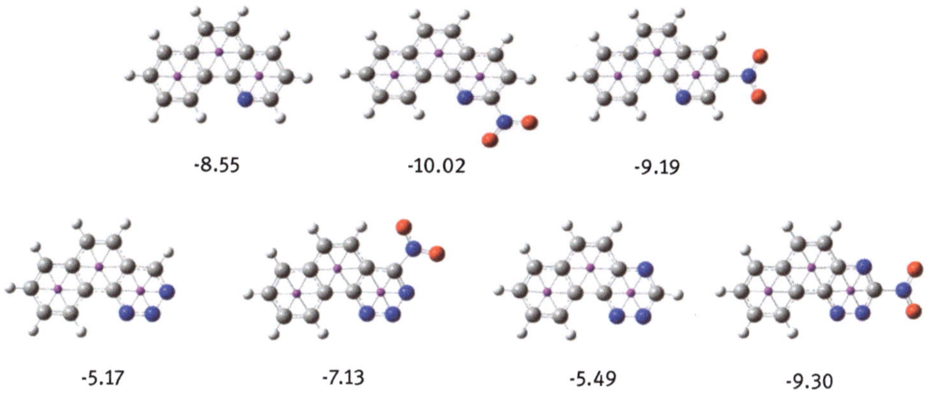

Figure 1.4: Effect of NO_2 substitutin on the aromaticity of the compounds. NICS data are in ppm.

Table 1.3: HOMA data for the present systems.

Structure	A Ring
Ph	0.898
Ph_1	0.785
Ph_2	0.768
Ph_3	0.754
Ph_4	0.775
Ph_5	0.880
Ph_1_2	0.716
Ph_1_3	0.704
Ph_1_4	0.721
Ph_1_5	0.788
Ph_2_3	0.707
Ph_2_4	0.712
Ph_2_5	0.795
Ph_3_4	0.712
Ph_3_5	0.783
Ph_4_5	0.780
Ph_1_2_3	0.584
Ph_1_2_4	0.569
Ph_1_2_3_4	0.538

1.7 Conclusion

In this chapter, aromatic characters of mono, di, tri and tetraaza substituted phenanthrene derivatives have been investigated theoretically by density functional theory applications with B3LYP/6-31 + G(d,p) method. As a result, the systems do not prefer the vicinal location of nitrogens. Secondly, the aromaticities depend on the position of the substituent. NICS and HOMA definitions of aromaticity, both reflected the same outcome in the present case. Therefore, either magnetic or geometric criteria of aromaticity definitions can be preferred for further studies. In addition, the effect position of the substituent has been investigated. The aromaticity of aza phenantherenes enhanced (possess greater negative NICS data) when the electron withdrawing atoms or groups are closer to the centrically substituted nitrogen atoms.

References

[1] Rabjohn N. Organic syntheses, collective, vol. 4. Washington, D.C.: The National Academies Press, 1963.
[2] Kalescky R, Kraka E, Cremer D. Description of Aromaticity with the help of vibrational spectroscopy: anthracene and phenanthrene. J Phys Chem A. 2014;118:223–37.
[3] Cheucci G. Hydrodehalogenation of halogenated pyridines and quinolines by sodium borohydride/N,N,NO,NO–tetramethylethylenediamine under palladium catalysis. Tet Lett. 2010;51:1562–5.
[4] Shestakov AN, Pankova AS, Kuznetsov MA. Cycloisomerization – a straightforward way to benzo[h]quinolines and benzo[c]acridines. Chem Heterocycl Compd. 2017;53:1103–13.
[5] Mamane V, Louerat F, Lehl J, Abboud MFY. A general and efficient method for the synthesis of benzo–(iso)quinoline derivatives. Tetrahedron. 2008;64:10699–705.
[6] Ball CJ, Gilmore J, Willis MC. Copper–catalyzed tandem C_N bond formation: an efficient annulative synthesis of functionalized cinnolines. Angew Chem Int Ed. 2012;51:5718–22.
[7] Bencini A, Lippolis V. 1,10–Phenanthroline: A versatile building block for the construction of ligands for various purposes. Coord Chem Rev. 2010;254:2096–180.
[8] Summers LA. The phenanthrolines. Adv Heterocycl Chem. 1978;22:1–69.
[9] Sammes PG, Yahioglu G. 1,10–phenanthroline: a versatile ligand. Chem Soc Rev. 1994;23:327–34.
[10] Luman CR, Castellano FN. Phenanthroline Ligands. In: McCleverty JA, Meyer TJ, Lever ABP, editors. Comprehensive coordination chemistry, vol. II. Oxford, UK: Elsevier, 2004:25.
[11] Minkin VI, Glukhovtsev MN, Simkin BY. Aromaticity and antiaromaticity:electronic and structural aspects. New York: Wiley, 1994.
[12] Schleyer PR, Jiao H. What is aromaticity? Pure Appl Chem. 1996;68:209–18.
[13] Glukhovtsev MN. Aromaticity today: energetic and structural criteria. J Chem Educ. 1997;74:132–6.
[14] Krygowski TM, Cyranski MK, Czarnocki Z, Hafelinger G, Katritzky AR. Aromaticity: a theoretical concept of immense practical importance. Tetrahedron. 2000;56:1783–96.
[15] Schleyer PR. Introduction: aromaticity. Chem Rev. 2001;101:1115–18.

[16] Cyranski MK, Krygowski TM, Katritzky AR, Schleyer PR. To what extent can aromaticity be defi ned uniquely? J Org Chem. 2002;67:1333–8.
[17] Schleyer PR, Maerker C, Dransfeld A, Jiao H, Hommes NJRE. Nucleus independent chemical shifts: a simple and effi cient aromaticity probe. J Am Chem Soc. 1996;118:6317–18.
[18] Jiao H, Schleyer PR. Aromaticity of pericyclic reaction transition structures: magnetic evidence. J Phys Org Chem. 1998;11:655–62.
[19] Schleyer PR, Kiran B, Simion DV, Sorensen TS. Does $Cr(CO)_3$ complexation reduce the aromaticity of benzene? J Amer Chem Soc. 2000;122:510–13.
[20] Quinonero D, Garau C, Frontera A, Ballaster P, Costa A, Deya PM. Quantifi cation of aromaticity in oxocarbons: the problem of the fi ctitious 'nonaromatic' reference system. Chem Eur J. 2002;8:433–8.
[21] Patchkovskii S, Thiel W. Nucleus–independent chemical shifts from semiempirical calculations. J Mol Model. 2002;6:67–75.
[22] Mohajeri A, Shahamirian M. Substituent effect on local aromaticity in mono and di–substituted heterocyclic analogs of naphthalene. J Phys Org Chem. 2010;23:440–50.
[23] Krygowski TM, Palusiak M, Płonka A, Zachara–Horeglad JE. Relationship between substituent effect and aromaticity – Part III: naphthalene as a transmitting moiety for substituent effect. J Phys Org Chem. 2007;20:297–306.
[24] Krygowski TM, Cyranski MK. Structural aspects of aromaticity. Chem Rev. 2001;101:1385–419.
[25] Gümüş S. The aromaticity of substituted diazanaphthalenes. Comput Theor Chem. 2011;963:263–7.
[26] Stewart JJP. Optimization of parameters for semi empirical methods I. Method J Comput Chem. 1989;10:209–20.
[27] Stewart JJP. Optimization of parameters for semi empirical methods II. Applications J Comput Chem. 1989;10:221–64.
[28] Leach AR. Molecular modelling. Essex: Longman, 1997.
[29] Kohn W, Sham LJ. Self–consistent equations including exchange and correlation effects. Phys Rev. 1965;140:1133–8.
[30] Parr RG, Yang W. Density functional theory of atoms and molecules. London: Oxford University Press, 1989.
[31] Becke AD. Density–functional exchange–energy approximation with correct asymptotic behavior. Phys Rev A. 1988;38:3098–100.
[32] Vosko SH, Vilk L, Nusair M. Accurate spin–dependent electron liquid correlation energies for local spin density calculations: a critical analysis. Can J Phys. 1980;58:1200–11.
[33] Lee C, Yang W, Parr RG. Development of the Colle–Salvetti correlation–energy formula into a functional of the electron density. Phys Rev B. 1988;37:785–9.
[34] Scuseria GE. Comparison of coupled–cluster results with a hybrid of Hartree–Fock and density functional theory. J Chem Phys. 1992;97:7528–30.
[35] Sosa C, Lee C. Density functional description of transition structures using nonlocal corrections. silylene insertion reactions into the hydrogen molecule. J Chem Phys. 1993;98:8004–11.
[36] Wilson PJ, Amos RD, Handy NC. Density functional predictions for metal and ligand nuclear shielding constants in diamagnetic closed–shell first–row transition–metal complexes. Phys Chem Chem Phys. 2000;2:187–94.
[37] Pulay P, Hinton JF, Wolinski K. Nuclear magnetic shieldings and molecular structure. In: Tossel JA, editor. NATO ASI series C, vol. 386. Netherlands: Kluwer, 1993.
[38] Hehre WJ, Radom L, Schleyer PR, Pople JA. Ab initio molecular orbital theory. New York: Wiley, 1986.

[39] Frisch MJ, Trucks GW, Schlegel HB, Scuseria GE, Robb MA, Cheeseman JR, et al. Wallingford, CT: Gaussian Inc., 2009.
[40] Yu L, Zhou X, Wu D, Xiang H. Synthesis of phenazines by Cu–catalyzed homocoupling of 2–halogen anilines in water. J Organometall Chem. 2012;705:75–8.
[41] Stanger A. What is . . . aromaticity: a critique of the concept of aromaticity—can it really be defined? Chem Commun. 2009;15:1939–47.

Nik Nur Syazana Bt Nik Mohamed Kamal, Theam Soon Lim,
Rusli Ismail and Yee Siew Choong

2 Conformations and interactions comparison between *R*- and *S*-methadone in wild type CYP2B6, 2D6 and 3A4

Abstract: Methadone is a morphine-substitute drug in methadone maintenance treatment (MMT) program to treat patients with opioid dependency. However, the methadone clinical effects are depending on the methadone metabolism rates that vary among the patients with genetic polymorphism of cytochrome P450s (CYPs). Our previous study showed methadone has different binding affinity due to the polymorphisms in CYP2B6, CYP2D6 and CYP3A4 that could contribute to the methadone metabolism rate. In this work, the conformation and interactions of *R*- and *S*-methadone in wild type CYP2B6, CYP2D6 and CYP3A4 were further studied in order to understand behaviour of *R*- and *S*-methadone at the CYP binding site. Clustering analysis showed that the conformation of *R*- and *S*-methadone in CYP2B6 are most stable, thus could lead to a higher efficiency of methadone metabolism. The conformation fluctuation of methadone in CYP2D6 could due to relatively smaller binding pocket compared with CYP2B6 and CYP3A4. The binding sites volumes of the studied CYPs were also found to be increased upon the binding with methadone. Therefore, this might contributed to the interactions of both *R*- and *S*-methadone in CYPs were mainly by hydrophobic contacts, van der Waals and electrostatic interactions. In the future, should an inhibitor for CYP is to be designed to prolong the prolonged opioid effect, the inhibitor should cater for single CYP isozyme as this study observed the behavioural differences of methadone in CYP isozymes.

Keywords: methadone, CYP2B6, CYP2D6, CYP3A4, molecular dynamics simulation

This article has previously been published in the journal Physical Sciences Reviews. Please cite as: Kamal, N. N. S. B. N. M., Lim, T. S., Ismail, R., Choong, Y. S. Conformations and interactions comparison between *R*- and *S*-methadone in wild type CYP2B6, 2D6 and 3A4 *Physical Sciences Reviews* [Online] 2019, 4 DOI: 10.1515/psr-2018-0172.

https://doi.org/10.1515/9783110631623-002

Graphical Abstract:

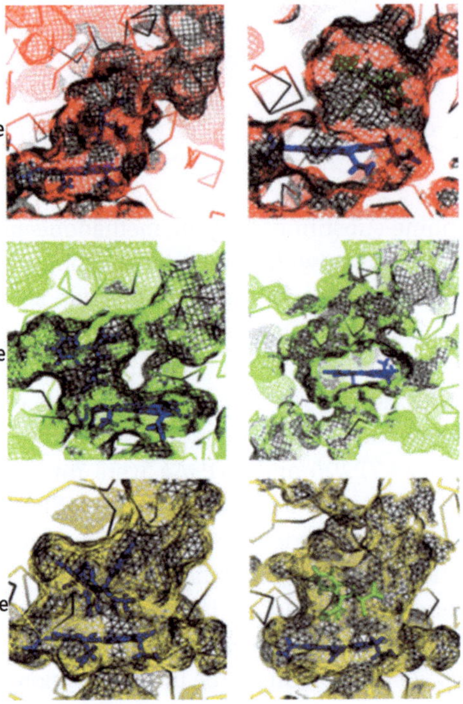

2.1 Introduction

Methadone (6-di**meth**yl**a**mino-4,4-**d**iphenyl-3-heptan**one)** is able to decrease the μ-receptor-related side effects with the lower affinity towards μ-receptor. On the other hand, the greater affinity towards δ-receptor (crucial for the development of morphine-induced tolerance and dependency) benefits the MMT program patients by relieving narcotic cravings [1–5]. However, inappropriate methadone dosage could also end up with fatalities or an increase in adverse events such as respiratory depression, prolongation QT interval, swelling of hands or feet and urinary retention [2, 6–9].

The methadone metabolism is by hepatic cytochrome P450 (CYP) enzymes including CYP2B6, CYP2D6 and CYP3A4 [10–14]. The single nucleotide polymorphisms (SNPs) of CYP are related to the differences in inter-individual variability in methadone metabolism rate, leads to differences in maximal plasma concentration, half-life and clearance of the drug. Free energy calculation from our previous docking studies further supported the above-mentioned phenomena where different binding affinity of *R*- and *S*-methadone in the alleles of CYP2B6, CYP2D6 and CYP3A4 were observed and that might lead to the differences in the methadone metabolism rate [15]. Thus, individual with difference methane metabolism rate will acquire different dosage in order to obtain the optimum end effects [13, 16, 17]. On the other hand, the marketed methadone are usually composed of racemix enantiomers (*R*- and *S*-methadone) with ratio of 1:1 [18]. The *R*-methadone is clinically significant μ-receptor agonist activity and has a longer terminal elimination half-life [19–21]. However, *S*-methadone is still important in delivering the pharmacology effects [22, 23].

The ability of above-mentioned docking simulation to predict the effects of CYP alleles on the binding affinity of methadone has thus driven the need to further study the interactions of methadone in the binding pocket of CYPs. Therefore, this work focused on the three main CYPs isozymes- CYP2B6, CYP2D6 and CYP3A4. We performed a relatively short molecular dynamics (MD) simulation for *R*- and *S*-methadone in complex with wild type CYP2B6, CYP2D6 and CYP3A4 to compare CYP-methadone complex. Results from the MD simulation showed that *S*-methadone is most stable in CYP2B6. The hydrophobic contacts and van der Waals interactions that drive the methadone-CYP complex formation might be due to the increased in CYPs binding pocket volume.

2.2 Methodology

The starting structure of *R*- and *S*-methadone were obtained from the lowest free energy in the most populated cluster conformation reported previously in our docking studies [15]. The 13 and 22 missing residues in CYP2D6 and CYP3A4, respectively, were added accordingly using MODELLER 9v10 [24]. A total of six systems were studied in this work and they were named according to the CYP (2B6, 2D6 and 3A4) and methadone enantiomers (*R*/*S*) i. e. 2B6-*R*, 2B6-*S*, 2D6-*R*, 2D6-*S*, 3A4-*R* and 3A4-*S*.

AMBER11 [25] was used in this MD simulation. The CYPs charges were first added using Amber FF03.r1 [26] while Amber GAFF [27] was applied for heme and methadone. Each system was solvated with 10 Å truncated octahedron TIP3P water model and neutralized with counterions. The minimization for the each system was was performed in two stages. Harmonic potential was used to restrain the solute during the first stage of minimization (1,000 steps steepest descent followed with 1,000 steps conjugate gradient). In the second stage of minimization, the whole system was minimized with 44,000 steps steepest descent and 14,000 steps conjugate gradient method without any restraint.

The solvent and ions was then heated from 0 K to 300 K by Langevin thermostat [28] in NVT ensemble with 10 Å cut-off with restraint on solute. The system was subsequently equilibrated without any restraint at 300 K using Langevin thermostat, 525,000 steps of 2 fs time step, NPT ensemble with average 1 atm pressure, 10 Å cut off, isotropic scaling and a relaxation time of 2 ps. The SHAKE algorithm [29] was turn on throughout the minimization process.

Production phase was divided into initial 100 ps of 5 kcal/mol/$Å^2$ restraints followed by 100 ps of 2 kcal/mol/$Å^2$ restraints, 10 Å cut-off, NVT ensemble and maintained at 300 K using Langevin dynamics. The system was then underwent 9.8 ns of simulation without any restraints in 10 Å cut-off, NPT ensemble, maintained at 300 K using Langevin thermostat; isotropic scaling and a relaxation time of 2 ps. The SHAKE algorithm was turn on throughout the process. The trajectories from the last 8 ns of the MD simulation were used for further analysis.

2.3 Results and discussion

Superimposition of CYP2B6, CYP2D6, and CYP3A4 crystal structures with the average MD structure did not show significant changes in the CYP backbone (Figure 2.1). This is further supported by root mean square deviations (RMSD) analysis of the systems. The RMSD plots for all systems were within 1.5–3.7 Å (Figure 2.2), thus showing that the systems are stable upon the interaction with methadone. In order to identify the region that contributed to the higher RMSD values, the analysis on the conformation of the system using root mean square fluctuation (RMSF) was performed. Results showed that the loop regions contributed to the highest fluctuation as expected (Figure 2.3). The highlighted region in Figure 2.3(d) – Figure 2.3(f) showed that

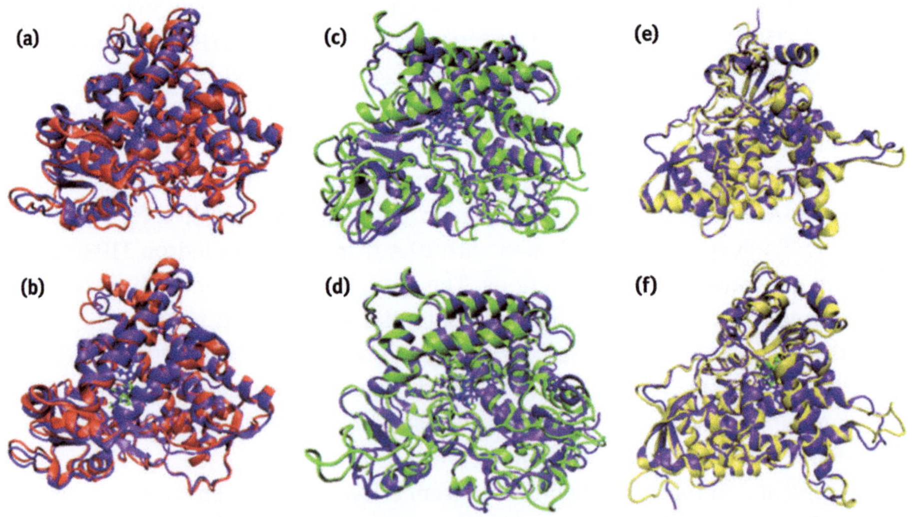

Figure 2.1: Superimposition of CYPs crystal structures (blue ribbon presentation) with the average MD structure (red, green and yellow ribbon for CYP2B6, CYP2D6 and CYP3A4, respectively, for (a) 2B6-*R*, (b) 2B6-*S*, (c) 2D6-*R*, (d) 2D6-*S*, (e) 3A4-*R*, and (f) 3A4-*S*. The methadone is in stick presentation.

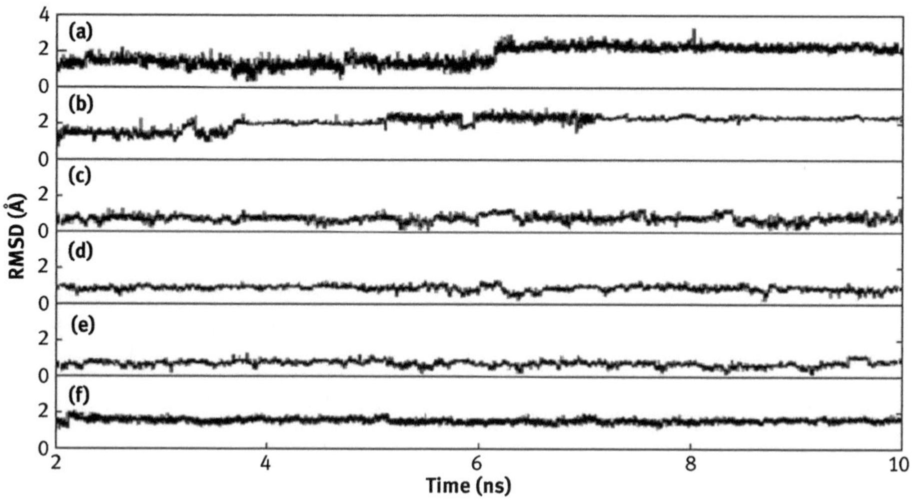

Figure 2.2: Root mean square deviation (RMSD) analysis of (a) 2B6-*R*, (b) 2B6-*S*, (c) 2D6-*R*, (d) 2D6-*S*, (e) 3A4-*R*, and (f) 3A4-*S*.

CYP2D6 Met374 and CYP3A4 Glu285 have fluctuation more than 3.0 Å. This might be due to the CYP3A4 Glu285 was the added missing residue (missing residues in the crystal structure; PDB id 3NXU) while Met347 is the located at the loop region in CYP3A4 and CYP2D6. The RMSD values of more than 8 Å recorded in 3A4-*R* and 3A4-*S* system are due to the C-terminal residues that are not involved in methadone binding.

The conformational changes of methadone were also studied using clustering analysis. A more stable conformation will result in a lower number of clusters. Clustering analysis of methadone showed a total of 6, 5, 34, 25, 10 and 9 clusters in 2B6-*R*, 2B6-*S*, 2D6-*R*, 2D6-*S*, 3A4-*R* and 3A4-*S* system, respectively (Figure 2.4). Therefore, *S*-methadone is most stable in CYP2B6, indicating preferable binding compared with other CYPs, supporting the experimental data that CYP2B6 preferentially metabolizes *S*-methadone [6, 12]. In addition, the more stable conformation of *R*- and *S*-methadone was found in CYP2B6 (lowest number of clusters compared to CYP2B6 and CYP3A4). This might suggest that the higher stability of methadone in CYP2B6 could result in higher affinity with CYP2B6 and thus lead to a higher methadone metabolism rate. This hypothesis is consistent with several reports that suggesting that CYP2B6 could be the major determinant of methadone metabolism [6, 14, 30–32]. On the other hand, highest number of clusters was observed for *R*- and *S*-methadone in CPY2D6 compared with CYP2B6 and CYP3A4. This result might due to the relatively smaller binding pocket of CYP2D6 compared to CYP2B6 and CYP3A4. Therefore, both *R*- and *S*-methadone were trying to search

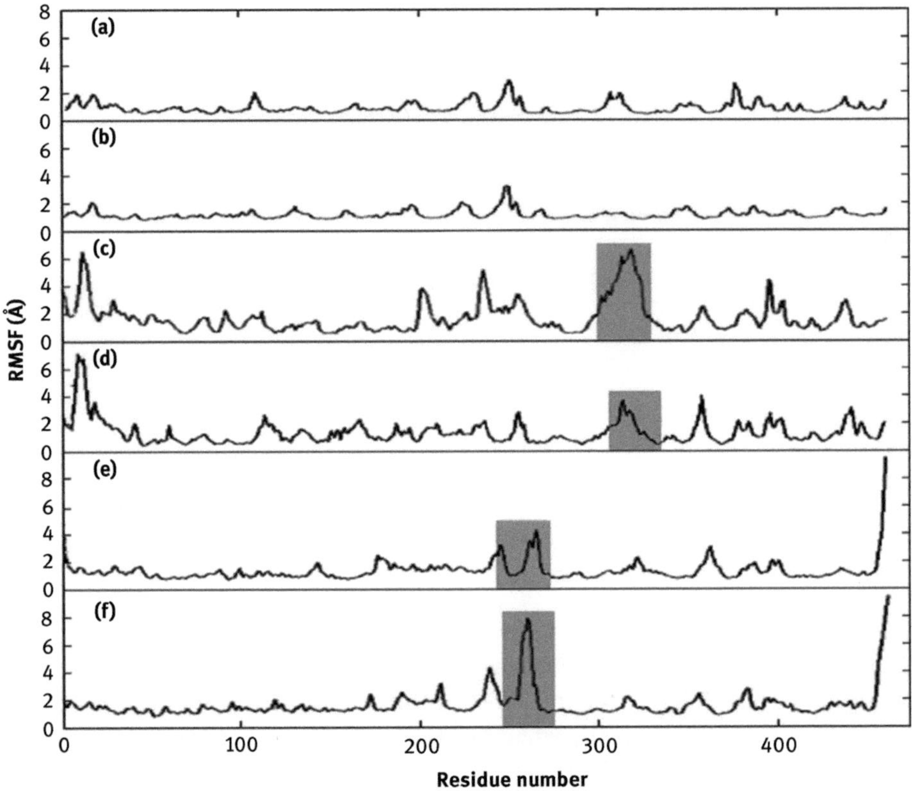

Figure 2.3: Root mean square fluctuation (RMSF) analysis for the CYPs in (a) 2B6-*R*, (b) 2B6-*S*, (c) 2D6-*R*, (d) 2D6-*S*, (e) 3A4-*R* and (f) 3A4-*S*. The highlighted regions are the residues with RMSF value of more than 3 Å.

for the best conformation to fit in the binding pocket. This finding is similar with the report whereby CYP2D is able to metabolize diverse drug classes, thus its binding site can sample different conformation [33].

The binding site of the CYP crystal structure was compared with the CYP average MD structure (Table 2.1 and Figure 2.5). The increase volume and size area but with the decrease of total number of pockets showed that changes could occur with the introduction of water molecules in the binding site. Only 2D6-*R* showed notable decrease in the area and volume size of the binding site. Therefore, this could also reason on the highest methadone confirmation fluctuation observed from clustering analysis for methadone in complex with CYP2D6. CYPs with *S*-methadone did show the influenced of water with increment of binding site compared to *R*-methadone. This might due to that *S*-methadone was docked and hold towards the binding site entrance of CYPs rather than more centralized *R*-methadone's location as reported earlier [15]. In 2B6-*R* and 2B6-*S*, the CYP binding sites were slightly wider with that of

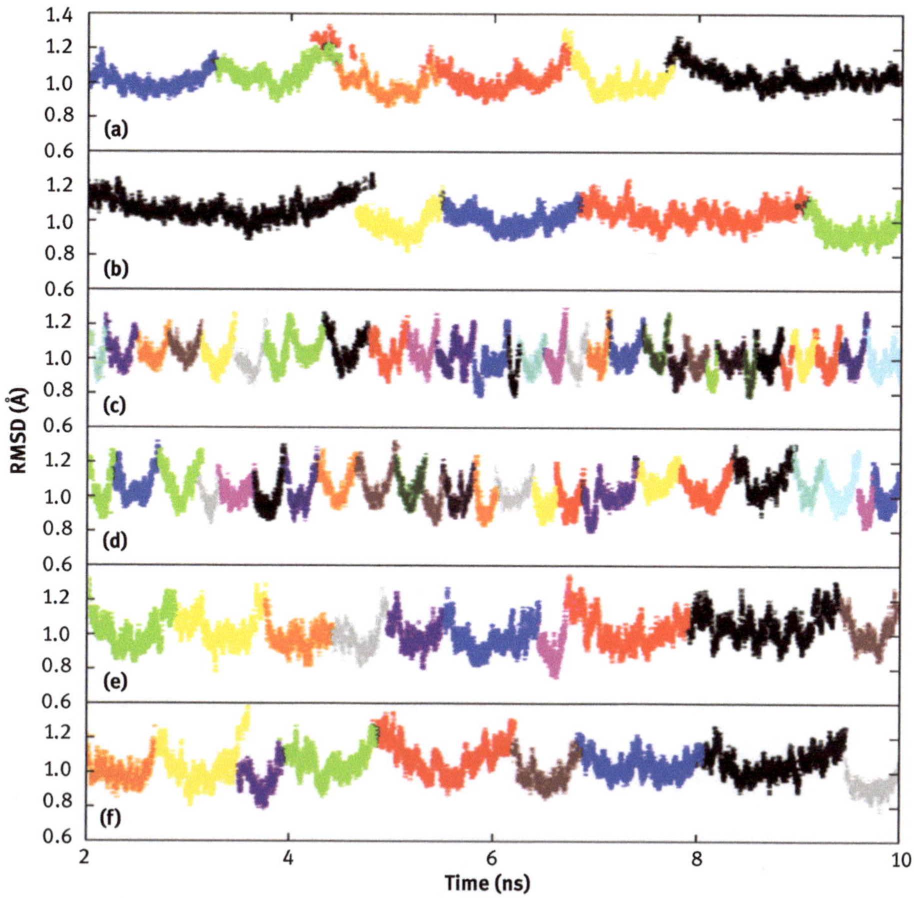

Figure 2.4: Clustering analysis of (a) 2B6-*R*, (b) 2B6-*S*, (c) 2D6-*R*, (d) 2D6-*S*, (e) 3A4-*R* and (f) 3A4-*S*.

crystal structure. As for CYP2D6 and CYP3A4, the binding sites were expanded compare with that of crystal structure. In addition, only 3A4-*R* has the increased in total number of pockets for average structure with *R*-methadone. Therefore, *R*-methadone could be affected by the introduction of water molecules in CYP3A4. The observation is similar with the reported study by Hendrychova et al. [34].

In general, a ligand would require some form of non-bonded interactions in order to be remained in a binding site. These interactions are often contributed by hydrogen bonding, ring stacking interactions, or weaker interactions such as hydrophobic contacts, van der Waals and electrostatic interactions [35]. However, hydrogen bond analysis did not show hydrogen bond formation between methadone and CYP for the occurrence of more than 5 % throughout MD simulation. This was supported by our earlier molecular docking simulation data where neither hydrogen bonding nor ring-stacking were observed between wild type CYP and methadone.

Table 2.1: The volume calculation for the binding site of CYP2B6, CYP2D6 and CYP3A4.

System	Number of pocket	Area (ms)	Area (sa)	Volume (ms)	Volume (sa)
Crystal CYP2B6	74	2029.69	1006.40	2499.99	426.72
2B6-*R*	68	2163.64	1204.03	3166.94	846.50
2B6-*S*	111	2052.27	1887.78	2838.07	746.83
Crystal CYP2D6	67	2083.19	1121.73	2989.11	824.35
2D6-*R*	54	1887.78	829.17	2183.60	390.46
2D6-*S*	72	6497.55	3873.36	9844.88	2332.22
Crystal CYP3A4	76	1785.43	1093.47	2866.66	891.40
3A4-*R*	85	2337.00	1374.37	3753.92	1232.62
3A4-*S*	96	2785.39	1540.80	4305.88	1410.89

sa = solvent accessible surface (SA, Richards' surface); ms = molecular surface (MS, Connolly's surface)

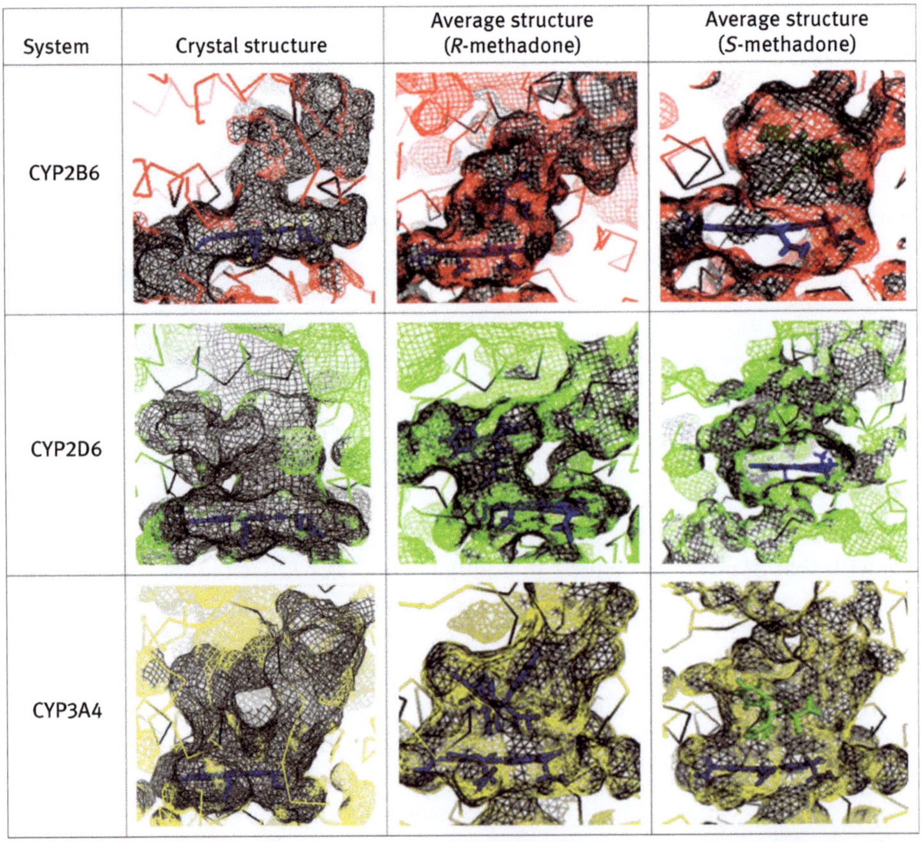

Figure 2.5: The binding site volume of CYP2B6, CYP2D6 and CYP3A4. Black pocket represents the calculated binding site. Heme is in blue stick presentation, centred in the binding site platform. The blue/green stick above the heme is *R*-methadone and *S*-methadone, respectively. Red, green and yellow lines are the binding residues of CYP2B6, CYP2D6 and CYP3A4, respectively.

Therefore, the driving force for CYP-methadone formation could be by the weaker non-bonded interactions, i. e. the hydrophobic contacts, van der Waals and electrostatic interactions, evidenced also from our previous analysis from docking simulation [15].

2.4 Conclusions

This work compared the interactions of *R*- and *S*-methadone in complex with three main cytochrome P450 isoenzymes, namely CYP2B6, CYP2D6 and CYP3A4. Overall structural stability was observed where methadone is most sable in CYP2B6 supporting the earlier predicted methadone binding site in CYPs by molecular docking simulation. The analysis from this work also consistent with other reported observations that suggest CYP2B6, followed with CYP3A4 could the major determinant for methadone metabolism. The conformation fluctuation of methadone in CYP2D6 could be due to the relatively smaller binding pocket compared with CYP2B6 and CYP3A4. On the other hand, hydrogen bonding was not observed between methadone and CYPs. This showed that the complex formation of methadone with CYPs is driven by weaker non-bonded interactions, i. e. hydrophobic contacts, van der Waals and electrostatic interactions. The CYPs binding site architecture showed that the introduction of water molecules affected the binding pocket size, especially CYPs in complex with *S*-methadone. The analysis in this work could be useful in the future should an inhibitor is to be designed in order to decrease methadone clearance to achieve the prolonged opioid effects. As behaviour differences between the isozymes were observed in this study by using one substrate (methadone), therefore the inhibitor should be designed to cater for only a single CYP isozyme.

Acknowledgements: This work was supported by Fundamental Research Grant Scheme (FRGS;203/CIPPM/6711680) from Malaysia Ministry of Education. Appreciation also extended to Universiti Sains Malaysia Postgraduate Research Scholarship for NNSBNM Kamal.

References

[1] Bowen S, Somohano VC, Rutkie RE, Manuel JA, Rehder KL. Mindfulness-based relapse prevention for methadone maintenance: A feasibility trial. J Altern Complement Med. 2017;23:541–4.
[2] Chhabra S, Bull J. Methadone. Am J Hosp Palliat Care. 2008;25:146–50.
[3] El Hage C, Ghabrash MF, Dubreucq S, Brissette S, Lesperance F, Lesperance P, et al. A pilot, open-label, 8-week study evaluating desvenlafaxine for treatment of major depression in methadone-maintained individuals with opioid use disorder. Int Clin Psychopharmacol. 2018;33:268–73.

[4] Fishman SM, Wilsey B, Mahajan G, Molina P. Methadone reincarnated: novel clinical applications with related concerns. Pain Med. 2002;3:339–48.
[5] Maremmani AG, Pacini M, Maremmani I. What we have learned from the methadone maintenance treatment of dual disorder heroin use disorder patients. Int J Environ Res Public Health. 2019;16:447.
[6] Gadel S, Crafford A, Regina K, Kharasch ED. Methadone N-demethylation by the common CYP2B6 allelic variant CYP2B6.6. Drug Metab Dispos. 2013;41:709–13.
[7] Kharasch ED, Hoffer C, Whittington D, Walker A, Bedynek PS. Methadone pharmacokinetics are independent of cytochrome P4503A (CYP3A) activity and gastrointestinal drug transport: insights from methadone interactions with ritonavir/indinavir. Anesthesiology. 2009;110:660–72.
[8] Peng S, Jiang H, Du J, Lin S, Pan S, Yu S, et al. Methadone dosage and plasma levels, SNPs of OPRM1 gene and age of first grug use were associated with outcomes of methadone maintenance treatment. Front Genet. 2018;9:450.
[9] Russolillo A, Moniruzzaman A, Somers JM. Methadone maintenance treatment and mortality in people with criminal convictions: A population-based retrospective cohort study from Canada. PLoS Med. 2018;15:e1002625.
[10] Fonseca F, de la Torre R, Diaz L, Pastor A, Cuyas E, Pizarro N, et al. Contribution of cytochrome P450 and ABCB1 genetic variability on methadone pharmacokinetics, dose requirements, and response. PLoS One. 2011;6:e19527.
[11] Kharasch ED, Hoffer C, Whittington D, Sheffels P. Role of hepatic and intestinal cytochrome P450 3A and 2B6 in the metabolism, disposition, and miotic effects of methadone. Clin Pharmacol Ther. 2004;76:250–69.
[12] Levran O, Peles E, Hamon S, Randesi M, Adelson M, Kreek MJ. CYP2B6 SNPs are associated with methadone dose required for effective treatment of opioid addiction. Addict Biol. 2013;18:709–16.
[13] Volpe DA, Xu Y, Sahajwalla CG, Younis IR, Patel V. Methadone metabolism and drug-drug interactions: in vitro and in vivo literature review. J Pharm Sci. 2018;107:2983–91.
[14] Kharasch ED. Current concepts in methadone metabolism and transport. Clin Pharmacol Drug Dev. 2017;6:125–34.
[15] Kamal NN, Lim TS, Tye GJ, Ismail R, Choong YS. The effect of CYP2B6, CYP2D6 and CYP3A4 alleles on methadone bonding: A molecular docking study. J Chem. 2013;2013:249642.
[16] Shiran MR, Lennard MS, Iqbal MZ, Lagundoye O, Seivewright N, Tucker GT, et al. Contribution of the activities of CYP3A, CYP2D6, CYP1A2 and other potential covariates to the disposition of methadone in patients undergoing methadone maintenance treatment. Br J Clin Pharmacol. 2009;67:29–37.
[17] Fonseca F, Torrens M. Pharmacogenetics of methadone response. Mol Diagn Ther. 2018;22:57–78.
[18] Maremmani I. The Principles and Practice of Methadone Treatment. Pacini Editore Medicina & AU–CNS. Pisa. 2009.
[19] Karch SB, Stephens BG. Toxicology and pathology of deaths related to methadone: retrospective review. West J Med. 2000;172:11–14.
[20] Foster DJ, Somogyi AA, Dyer KR, White JM, Bochner F. Steady-State Pharmacokinetics of (R)- and (S)-Methadone in Methadone Maintenance Patients. Clin Pharmacol. 2000;50:427–40.
[21] Kharasch ED, Regina KJ, Blood J, Friedel C. Methadone pharmacogenetics: CYP2B6 polymorphisms determine plasma concentrations, clearance, and metabolism. Anesthesiology. 2016;123:1142–53.
[22] Ahmad T, Valentovic MA, Rankin GO. Effects of cytochrome P450 single nucleotide polymorphisms on methadone metabolism and pharmacodynamics. Biochem Pharmacol. 2018;153:196–204.

[23] Peng PW, Tumber PS, Gourlay D. Perioperative Pain Management of Patients on Methadone Therapy. Reg Anesthesia Pain. 2005;52:513–23.
[24] Sali A, Blundell TL. Comparative protein modelling by satisfaction of spatial restraints. J Mol Biol. 1993;234:779–815.
[25] Case DA, Darden TA, Cheatham IT, Simmerling CL, Wang J, Duke RE, et al. AMBER11. San Franscisco: University of California, 2010.
[26] Duan Y, Wu C, Chowdhury S, Lee MC, Xiong G, Zhang W, et al. A point-charge force field for molecular mechanics simulations of proteins based on condensed-phase quantum mechanical calculations. J Comput Chem. 2003;24:1999–2012.
[27] Wang J, Wolf RM, Caldwell JW, Kollman PA, Case DA. Development and testing of a general amber force field. J Comput Chem. 2004;25:1157–74.
[28] Izaguirre JA, Catarello DP, Wozniak JM, Skeel RD. Langevin stabilization of molecular dynamics. J Chem Phys. 2001;114:2090–8.
[29] Ryckaert J-P, Ciccotti G, Berendsen HJ. Numerical integration of the cartesian equations of motion of a system with constraints: molecular dynamics of n-alkanes. J Comput Phys. 1977;23:327–41.
[30] Chang Y, Fang WB, Lin SN, Moody DE. Stereo-selective metabolism of methadone by human liver microsomes and cDNA-expressed cytochrome P450s: A reconciliation. Basic Clin Pharmacol Toxicol. 2011;108:55–62.
[31] Totah RA, Allen KE, Sheffels P, Whittington D, Kharasch ED. Enantiomeric metabolic interactions and stereoselective human methadone metabolism. J Pharmacol Exp Ther. 2007;321:389–99.
[32] Totah RA, Sheffels P, Roberts T, Whittington D, Thummel K, Kharasch ED. Role of CYP2B6 in stereoselective human methadone metabolism. Anesthesiology. 2008;108:363–74.
[33] Olubiyi OO, Olagunju MO, Obisesan AO. Computational analysis of physicochemical factors driving CYP2D6 ligand interaction. Curr Comput Aided Drug Des. 2017;13:39–47.
[34] Hendrychova T, Anzenbacherova E, Hudecek J, Skopalik J, Lange R, Hildebrandt P, et al. Flexibility of human cytochrome P450 enzymes: molecular dynamics and spectroscopy reveal important function-related variations. Biochim Biophys Acta. 2011;1814:58–68.
[35] Sano E, Li W, Yuki H, Liu X, Furihata T, Kobayashi K, et al. Mechanism of the decrease in catalytic activity of human cytochrome P450 2C9 polymorphic variants investigated by computational analysis. J Comput Chem. 2010;31:2746–58.

Aleksey E. Kuznetsov

3 Phthalocyanines core-modified by P and S and their complexes with fullerene C_{60}: DFT study

Abstract: Phthalocyanines (Pcs) and their derivatives have attracted a lot of attention because of their both biological importance and technological applications. The properties of Pcs can be tuned by replacing the central atom, by modifying the periphery of phthalocyanine ring, and by changing the *meso*-atoms. One more promising pathway for modifying Pcs and their derivatives can be the *core-modification*, or substitution of the core isoindole nitrogen(s) by other elements. Motivated by the results obtained for some core-modified porphyrins, we investigated computationally complete core-modification of regular Zn phthalocyanine (ZnPc) with P and S. We performed density functional theory studies of the structures, charges, and frontier molecular orbitals of P-core-modified and S-core-modified ZnPcs, $ZnPc(P)_4$ and $ZnPc(S)_4$, using both B3LYP and two dispersion-corrected functionals. Also, we studied computationally formation of complexes between the fullerene C_{60} and $ZnPc(P)_4$ and $ZnPc(S)_4$. Both $ZnPc(P)_4$ and $ZnPc(S)_4$ show strong bowl-like distortions similar to the results obtained earlier for $ZnP(P)_4$ and $ZnP(S)_4$. The size of the "bowl" cavity of the both core-modified Pcs is essentially the same, showing no dependence on the core-modifying element. For $ZnPc(S)_4$, the HOMO is quite different from those of ZnPc and $ZnPc(P)_4$. When the fullerene C_{60} forms complexes with the $ZnPc(P)_4$ and $ZnPc(S)_4$ species in the gas phase, it is located relatively far (4.30–5.72 Å) from the one of the P-centers and from the Zn-center of $ZnPc(P)_4$, whereas with $ZnPc(S)_4$ C_{60} forms relatively short bonds with the Zn-center, varying from ca. 2.0 to ca. 3.0 Å. The very strong deformations of both the $ZnPc(P)_4$ and $ZnPc(S)_4$ structures are observed. The calculated binding energy at the B3LYP/6-31G* level for the C_{60}-$ZnPc(P)_4$ complex is quite low, 1.2 kcal/mol, which agrees with the quite long distances fullerene - $ZnPc(P)_4$, whereas it is noticeably larger, 13.6 kcal/mol, for the C_{60}-$ZnPc(S)_4$ complex which again agrees with the structural features of this complex. The binding energies of the complexes optimized using the dispersion-corrected functionals, CAM-B3LYP and wB97XD, are significantly larger, varying from ca. 14 till 52 kcal/mol which corresponds with the shorter distances between the fullerene and $ZnPc(X)_4$ species.

Keywords: phthalocyanines, core-modification, DFT, fullerene

This article has previously been published in the journal Physical Sciences Reviews. Please cite as: Kuznetsov, A. E. Phthalocyanines core-modified by P and S and their complexes with fullerene C60: DFT study *Physical Sciences Reviews* [Online] 2019, 4 DOI: 10.1515/psr-2019-0001.

https://doi.org/10.1515/9783110631623-003

3.1 Introduction

Phthalocyanines (Pcs) and their derivatives have caused a lot of interest and attracted much attention because of their both biological importance and technological applications [1–3]. Generally speaking, the Pcs and their derivatives have the following characteristic features [4–9]: (1) a special two-dimensional conjugated 18π-electron structure [1]; (2) high stability towards light and heat; (3) their molecular structures are diverse and easy to tailor, they can be modified by a variety of ways, and (4) the Pcs coordination ability is very strong. This ability to bind a variety of transition metals (among other elements) and easily tailorable peripheral substitution are responsible for the Pcs outstanding redox properties [10], singlet oxygen generation [11], conductivity [12] and other properties [13]. Chemical, optical, coordination, electrochemical, and other properties of Pcs can be easily tuned employing the three major pathways: (i) by replacing the central atom (see, e. g. [6–14]); (ii) by modifying the periphery of Pc ring (see, e. g. [5, 7, 13, 15–19]) and (iii) by changing the *meso*-atoms.

Due to their extensive coordination chemistry along with other important characteristics, such as large absorption coefficients in the visible region and high thermal and photochemical stability [1–3, 7], Pcs and their derivatives have been extensively employed in a number of applications: chemical sensors [1, 10, 20], solar cells [21–29], catalysts [1, 2, 7, 30–34], electrophotography [1, 3], optoelectronic materials [35, 36], liquid crystals [37], Langmuir-Blodgett films [38], nanotechnology [39–42], photosensitizers in photodynamic cancer therapy [16, 43–47], two-dimensional mesoporous polymers and covalent organic frameworks [48], etc. One more promising pathway for modifying structures and properties of Pcs and their derivatives can be the so-called *core-modification*, or substitution of the core isoindole nitrogen(s) by other elements (see, e. g. [49–53]). Recently, we reported the computational studies of the series of metalloporphyrins (MPs) with all the four pyrrole nitrogens replaced with P-atoms, $MP(P)_4$, for M = Sc-Zn [54–59]. We demonstrated that the prominent structural feature of all the $MP(P)_4$ compounds under investigation is their significant distortion from planarity leading to their bowl-like shapes [54–59]. Furthermore, motivated by the multiple examples of formation of stacks by regular MPs and their derivatives, we performed the computational studies of the stack formation between the $ZnP(P)_4$ species without any linkers or substituents [57]. We found the three possible modes of binding or coordination between the monomeric $ZnP(P)_4$ units, with the so-called "convexity-to-convexity" dimer being the most stable compound among the three types of dimers studied. Next, motivated by the numerous examples of the complexation between regular planar or quasi-planar MPs and fullerene C_{60}, we computationally investigated possibility of the complex formation between $ZnP(P)_4$ and $NiP(P)_4$ and C_{60} without any linkers, using the CAM-B3LYP/6-31G* approach, both in the gas phase and with implicit effects from C_6H_6 [58]. We found that the $MP(P)_4$-C_{60} complexes could indeed form, with the binding energies being relatively low, ca. 1–1.6 kcal/mol and ca. 5 kcal/mol for M = Zn and Ni, respectively. Also, it was found that the $ZnP(P)_4$ species is noticeably distorted in the $ZnP(P)_4$-C_{60} complex whereas $NiP(P)_4$ inside the $NiP(P)_4$-C_{60} complex essentially retained its bowl-like shape. Very recently, motivated by the numerous examples of the formation of complexes/nanoassemblies between various nanoparticles (NPs) and regular MPs and their derivatives and relative scarcity of their computational studies, we decided to investigate computationally if the complex formation between the core-modified $MP(X)_4$ porphyrins and semiconductor quantum dots, exemplified by small NP Zn_6S_6, without any substituents or linkers would be possible [59]. For this study we chose two core-modified Zn-porphyrins, $ZnP(P)_4$ and $ZnP(S)_4$. We decided to focus on these species because they are relatively simple representatives of the core-modified $MP(X)_4$ compounds, with relatively "inactive" d-electrons. The complexes formation was investigated using two theoretical approaches: (i) B3LYP/6-31G* and (ii) CAM-B3LYP/6-31G*, both in the gas phase and with implicit effects from C_6H_6 considered. The calculated binding energies of the complexes studied were found to be significant, varying from ca. 29 up to ca. 69 kcal/mol, depending on the complex and the approach employed. The core-modified porphyrin species were found to become noticeably distorted upon the complex formation, although not as strong as in the case of $ZnP(P)_4$-C_{60} complex [59]. Some charge transfer was found to occur both from the $ZnP(X)_4$ porphyrin macrocycles to the Zn_6S_6 NP and within the $ZnP(X)_4$ porphyrin macrocycles and the NP itself.

Thus, motivated by the above-described results obtained for some core-modified porphyrins and their complexes, we decided to investigate computationally what would occur upon complete core-modification of regular Zn phthalocyanine (ZnPc) with P and S. We performed DFT studies of the structures, charges, and frontier molecular orbitals of P-core-modified and S-core-modified ZnPcs, $ZnPc(P)_4$ and $ZnPc(S)_4$, respectively, and compared them with the regular ZnPc. Also, we studied computationally formation of complexes between the fullerene C_{60} and $ZnPc(P)_4$ and $ZnPc(S)_4$ (see below). The paper is organized as follows: in the following section, we address the computational details of the study; next, we consider structural features, charges and frontier MOs of ZnPc, $ZnPc(P)_4$ and $ZnPc(S)_4$; next, we address the C_{60}-$ZnPc(X)_4$ (X = P, S) complexes; finally, we summarize the research findings and discuss further research perspectives.

3.2 Computational details

The study described here was performed using the Gaussian 09 package [60]. The ZnPc, $ZnPc(P)_4$ and $ZnPc(S)_4$ species were optimized without any symmetry constraints, and the resulting structures were assessed using vibrational frequency analysis to probe whether or not they represent true minimum-energy geometries.

We performed the geometry optimizations and frequencies calculations using the hybrid functional B3LYP [61] with the split-valence polarized 6-31G* basis set [62–66], furthermore referred to as B3LYP/6-31G* approach. Earlier, the B3LYP method with the 6-31G* basis set was proved to give geometries in good agreement with experiments (see, e. g. [67]), and was shown to produce the ordering of spin states of metalloporphyrin complexes reasonably well [68]. The DFT approaches have been successfully employed in computational studies of both Pcs (see, e. g. [13–17, 22, 48–50, 69–71]) and their complexes with the fullerene C_{60} [26, 72–80]. To study the C_{60}-ZnPc(X)$_4$ (X = P, S) complexes we first employed the B3LYP/6-31G* approach and then the Handy and coworkers' long range corrected version of B3LYP using the Coulomb-attenuating method, CAM-B3LYP [81] along with the wB97XD functional which uses a version of Grimme's D2 dispersion model [82], with the same basis set. Using these functionals to study the complexes under discussion is very important due to a crucial role of van der Waals (vdW) forces in the interactions between Pc and fullerene units [72]. To speed up the calculations, the complex structures along with their components were also optimized at the B3LYP/3-21G* [83, 84] level of theory.

The binding energies (E_{bind}) of the C_{60}-ZnPc(X)$_4$ complexes were computed using the following formula:

$$E_{bind} = E\left(C_{60} - ZnPc(X)_4\right) - E(C_{60}) - E\left(ZnPc(X)_4\right).$$

Below we consider the gas-phase results without the zero-point correction ZPE (ΔE_0). The charge analysis was performed using the Natural Bond Orbital (NBO) scheme with the "pop = nbo" command as implemented in the Gaussian 09 package [85, 86]. Molecular structures and MOs were visualized using OpenGL version of Molden 5.0 visualization software [87].

3.3 Results and discussion

3.3.1 Comparison of structural features of ZnPc, ZnPc(P)$_4$ and ZnPc(S)$_4$

For the two of the three Pc species studied, ZnPc and ZnPc(S)$_4$, singlet structures were found to be the lowest in energy at all the computational approaches employed (cf. Table S1, Supporting Information). Thus, at the B3LYP/6-31G* (B3LYP/3-21G*)//CAM-B3LYP/6-31G*//wB97XD/6-31G* levels, the triplet-singlet energy differences were calculated to be quite noticeable: ZnPc, 25.8 (20.1)//20.3//18.1; ZnPc(S)$_4$, 17.8 (10.4)//11.8//10.9 kcal/mol, respectively. However, the situation was found to be different for the ZnPc(P)$_4$ compound: at the B3LYP/6-31G* (B3LYP/3-21G*) the triplet-singlet energy differences were again calculated to be quite noticeable, 10.1 (11.0) kcal/mol. But with the approaches taking the dispersion corrections in the account, the singlet and triplet structures were found to be very close to each other:

the singlet-triplet difference calculated with the wB97XD/6-31G* approach was mere 1.5 kcal/mol, and with the CAM-B3LYP/6-31G* approach the triplet structure was calculated to be mere 0.04 kcal/mol lower than the singlet. Thus, for the heavier congener of N, phosphorus, the triplet structure becomes less energetically unfavorable (cf. 10.1 vs. 25.8 kcal/mol for ZnPc) or even energetically comparable with the singlet structure. (Interestingly, for the next heavier congener, As, we found the triplet structure at the B3LYP/6-31G* level to be even closer to the singlet, 5.7 kcal/mol). Interesting, for the core-modified porphyrins $ZnP(X)_4$ studied using the same B3LYP/6-31G* approach, the singlet-triplet energy differences were found to be 17.6 kcal/mol [55] and −3.6 [59] kcal/mol for X = P and S, respectively. That is, extending the carbon framework of the completely core-modified porphyrinic species may cause changes in their ground spin states (cf. Table S1).

As for the structural features of the three Pc species under investigation, for the completely core-modified ZnPcs we can see the drastic differences compared with the essentially flat ZnPc (cf. Figure 3.1 and Table 3.1) (it should be noticed that the B3LYP/3-21G* approach gave some deviations from planarity for this compound which could be ascribed to not very effective description of binding within this species with smaller basis set). As can be seen, both $ZnPc(P)_4$ and $ZnPc(S)_4$ show strong bowl-like distortions similar to the results obtained earlier for $ZnP(P)_4$ [54–59] and $ZnP(S)_4$ [59]. We took the following structural parameters as qualitative measures of

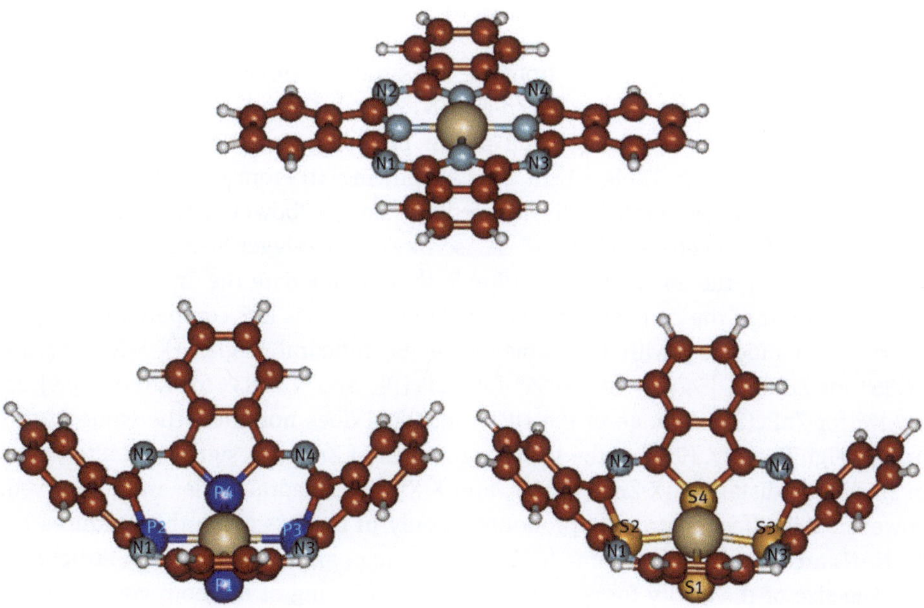

Figure 3.1: Singlet structures of the Pcs studied: ZnPc (top), $ZnPc(P)_4$ (bottom left) and $ZnPc(S)_4$ (bottom right). Color coding: dark brown for C, light grey for H, light brown for Zn, light blue for N, dark blue for P, dark yellow for S.

Table 3.1: Selected structural parameters of the ZnPc, $ZnPc(P)_4$ and $ZnPc(S)_4$ calculated at the following levels of theory in the gas phase: B3LYP/6-31G* (B3LYP/3-21G*) // CAM-B3LYP/6-31G* // wB97XD/6-31G*.

Molecule	$R(Zn\text{-}X)_{av}$, Å (X = N, P, S)	$\angle(X\text{-}Zn\text{-}X)_{av}$, ° (X = N, P, S)	$\angle(X\text{-}X\text{-}X\text{-}Zn)_{av}$, °(X = N, P, S)	$\angle(N_{1,2}\text{-}Zn\text{-}N_{4,3})_{av}$, °	$R(N_1\text{-}N_4)/R(N_2\text{-}N_3)$, Å
ZnPc	1.99 (1.97) // 1.98 // 1.98	178.46 (164.74) // 179.46 // 179.18	0.0 (10.71) // 0.38 // 0.58	178.95 (169.28) // 179.60 // 179.36	–
$ZnPc(P)_4$	2.34 (2.32) // 2.32 (2.29, 2.38)[a] // 2.32 (2.29, 2.39)[a]	179.51 (163.63) // 178.76 (177.32, 171.67)[a] // 179.63 (177.89, 170.38)[a]	−0.35 (−11.50) // −0.87 (1.93, −5.77)[a] // −0.26 (1.51, −6.65)[a]	137.33 (147.69) // 136.96 (137.51)[a] // 137.77 (137.79)[a]	7.08 (7.09) // 7.05 (7.03)[a] // 7.05 (7.02)[a]
$ZnPc(S)_4$	2.33 (2.34) // 2.31 // 2.32	162.65 (147.16) // 161.65 // 159.17	−12.17 (−22.52) // −12.86// −14.57	151.10 (162.35) // 152.80 // 154.40	7.08 (7.10) // 7.06 // 7.06

[a]Triplet structure

distortion of the $ZnPc(X)_4$ species from the flat structure: angles $(X\text{-}Zn\text{-}X)_{av}$, dihedral angles $(X\text{-}X\text{-}X\text{-}Zn)_{av}$, and angles $(N_1\text{-}Zn\text{-}N_4)/(N_2\text{-}Zn\text{-}N_3)$ (these last angles are equivalent to the parameter C_m-Zn-C_m' used in our previous study [55]). Analysis of the structural parameters in Table 3.1 shows the following: (i) From $ZnPc(P)_4$ to $ZnPc(S)_4$, the Zn-center becomes significantly "dented" inside the "bowl cavity" formed by the core-modified Pc moiety which could be ascribed to the bigger S size compared to P and thus making the S4-cavity less suitable to accommodate the Zn-center. (ii) The value of upshift of the Zn-center in the core-modified Pcs is less compared to the porphyrins core-modified with the same elements: dihedral angles $(X\text{-}X\text{-}X\text{-}Zn)_{av}$ are −4.36° for $ZnP(P)_4$ [54, 55] vs. −0.35° for $ZnPc(P)_4$ and −44.83° for $ZnP(S)_4$ [59] vs. −12.17° for $ZnPc(S)_4$. Change of the DFT functional does not affect the value of this upshift significantly. (iii) For the triplet structures of $ZnPc(P)_4$ significant alterations in Zn-X bond distances, X-Zn-X angles, and X-X-X-Zn dihedral angles were observed, however, this effect is strongly pronounced only in the P4-cavity: the distances N_1-N_4/N_2-N_3 are essentially the same for both the singlet and triplet $ZnPc(P)_4$ structures. (iv) The size of the cavity formed by the porphyrinic ring of the both core-modified Pcs, as judged by the distances N_1-N_4/N_2-N_3, is essentially the same (cf. distances R $(N_1\text{-}N_4)/R(N_2\text{-}N_3)$, Table 3.1), showing no dependence on the core-modifying element. (v) Using the B3LYP/3-21G* approach further shifts the Zn-center inside the "bowl cavities", however, the size of the cavity defined by the distances N_1-N_4/N_2-N_3 remains essentially unchanged. However, due to not very effective description of binding within this species with smaller basis set these results should be taken with caution. Using the dispersion-corrected functionals CAM-B3LYP and wB97XD was shown not to have significant effects on the calculated structural parameters of the compounds studied.

3.3.2 Electronic features of ZnPc, $ZnPc(P)_4$ and $ZnPc(S)_4$

Comparative analysis of the frontier MOs (see Figure 3.2) and selected NBO charges (see Table 3.2) of the ZnPc, $ZnPc(P)_4$ and $ZnPc(S)_4$ species shows the following (NBO charges are presented only at one level of theory because they do not change significantly with the method used). (i) There occurs reversal of the charge sign on the X-centers directly connected to the Zn-center for X = P and S, due to much smaller electronegativies of P (2.19) and S (2.58) compared to N (3.04). This was earlier observed for the P- and S-core-modified porphyrins [54–59]. (ii) Noticeable increase of the positive charge on Zn and decrease of the positive charge on X-centers from X = P to S. In general, we can suppose the accumulation of positive charge in the "bottom" part of the core-modified Pcs "bowl" to take place. (iii) From ZnPc to $ZnPc(P)_4$ to $ZnPc(S)_4$, first slight stabilization and then noticeable destabilization of HOMO is observed along with strong stabilization and then even stronger destabilization of LUMO (similar results obtained for $MP(P)_4$ [54, 55]). (iv) This, in turn, from ZnPc to $ZnPc(P)_4$ to $ZnPc(S)_4$ causes first strong decrease and then slight increase of the calculated HOMO-LUMO gap values (at the B3LYP/6-31G* level). However, for both the core-modified $ZnPc(X)_4$ the HOMO-LUMO gaps are smaller compared to the ZnPc, by 0.62 (X = P) and 0.29 (X = S) eV. (v) Similar trends in the HOMO-LUMO gap energies are observed with two other DFT functionals, although it should be noticed that the CAM-B3LYP and wB97XD functionals generally give significantly larger HOMO-LUMO gap values, compared to the B3LYP (by ca. 1.5–2.5 eV; that is why we visualized the frontier MOs of the species studied only at the B3LYP/6-31G* level). We understand that in general HOMO-LUMO gap values should be used with caution, and thus we performed TDDFT calculations with all three approaches employed. The obtained TDDFT gap values generally support the trends obtained for the HOMO-LUMO gap values. The only difference is for the triplet $ZnPc(X)_4$ TDDFT gap calculated at the wB97XD/6-31G* level: its value is essentially the same as for the wB97XD/6-31G* TDDFT gap for ZnPc. (vi) For both ZnPc and $ZnPc(P)_4$ HOMOs are quite similar in shape and in qualitative composition: no visible contributions from Zn, core N/P, and *meso* N (Figure 3.2). The LUMOs of these two species, however, have noticeable differences. As for $ZnPc(S)_4$, its HOMO is quite different from those of ZnPc and $ZnPc(P)_4$: some contributions from S- and *meso* N-centers become visible, along with different combinations of orbitals located at the isoindole units.

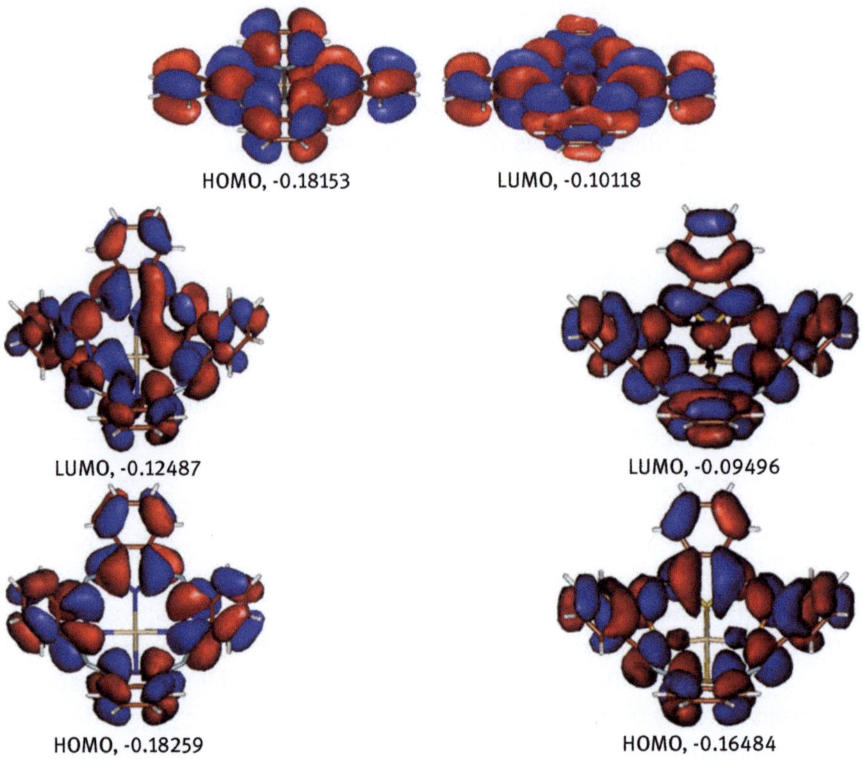

Figure 3.2: Frontier MOs of ZnPc (top), $ZnPc(P)_4$ (bottom left) and $ZnPc(S)_4$ (bottom right) calculated at B3LYP/6-31G* level.

Table 3.2: HOMO-LUMO gaps and TDDFT gaps, eV, calculated at the B3LYP/6-31G* // CAM-B3LYP/6-31G* // wB97XD/6-31G* level, and selected NBO charges, e, calculated at the B3LYP/6-31G* level,[b] for ZnPc, $ZnPc(P)_4$ and $ZnPc(S)_4$.

Species	HOMO-LUMO gap [TDDFT gap], eV	NBO charges, e	
		Zn	X
ZnPc	2.19 [2.09] // 3.75 [2.05] // 4.70 [2.03]	1.27	−0.69
$ZnPc(P)_4$	1.57 [1.20] // 3.03 [1.18] // 4.50,4.64 [2.02][a]	0.57	0.39
$ZnPc(S)_4$	1.90 [1.62] // 3.47 [1.61] // 4.41 [1.57]	0.91	0.19

[a]Triplet structure

The LUMO of this species has some similarities to the $ZnPc(P)_4$ LUMO showing some contributions from *meso* N-centers, along with slight contributions from the Zn-center and with different combinations of orbitals located at the isoindole units.

Such differences in the energetics and compositions of the frontier MOs of $ZnPc(P)_4$ and $ZnPc(S)_4$ compared to ZnPc might lead to different optical/photophysical properties of the core-modified compounds along with their different reactivities. Thus, we can suppose that the destabilized $ZnPc(S)_4$ LUMO would be more available to accept electron density from electron donor (e. g. to the Zn-center orbitals), as well as the destabilized $ZnPc(S)_4$ HOMO would be more available to donate electron density. The accumulation of positive charge in the "bottom" part of the core-modified Pcs "bowl", along with their strongly pronounced bowl-like deformation, could also lead to their different reactivity and charge transfer properties. More detailed analysis of the molecular orbitals, aromaticity, reactivities and charge transfer properties of these species is definitely necessary and will be the subject of the follow-up studies. Furthermore, in this study we decided to check the ability of these core-modified Pcs to make non-covalent complexes with the C_{60} fullerene.

It should be also noticed that so far there are no experimental or computational studies of completely core-modified metallophthalocyanines, thus our research could be considered as a pioneering study in this area.

3.3.3 Complexes C_{60}-$ZnPc(P)_4$ and C_{60}-$ZnPc(S)_4$

Intrigued by the strongly pronounced bowl-like shapes of the $ZnPc(P)_4$ and $ZnPc(S)_4$ compounds and motivated by the formation of the non-covalent C_{60}-$MP(P)_4$ (M = Zn, Ni) complexes studied before [58], we decided to check if the non-covalent complex formation would be possible between C_{60} and $ZnPc(P)_4$ and $ZnPc(S)_4$ compounds. To study these complexes, we employed first the B3LYP/3-21G* approach in the gas phase, and then refined the geometry using the B3LYP/6-31G*, CAM-B3LYP/6-31G* and wB97XD/6-31G* approaches, all in the gas phase (see Computational Details section). We understand that the results obtained with the B3LYP/3-21G* approach for complexes should be treated with some caution, thus we consider here the results obtained at the B3LYP/6-31G* level of theory, and the results obtained with other three approaches are given in the Supporting Information.

First of all, it is interesting to note that for the C_{60}-$ZnPc(P)_4$ complex both with the B3LYP/6-31G* and B3LYP/3-21G* approaches the singlet structure was calculated to be the lowest in energy, by 8.9 and 7.7 kcal/mol, respectively, whereas with the CAM-B3LYP/6-31G* and wB97XD/6-31G* approaches the triplet structure became more energetically favorable, by 5.9 and 4.4 kcal/mol, respectively (see Supporting Information). However, for the C_{60}-$ZnPc(S)_4$ complex situation is different: the triplet structure is the lowest-lying one with the B3LYP/6-31G*, B3LYP/3-21G*, and CAM-B3LYP/6-31G* approaches, being more stable by 4.4, 1.6 and 11.6 kcal/mol, respectively, but with the wB97XD/6-31G* approach the singlet becomes lower in energy by mere 0.2 kcal/mol (Supporting Information). Thus, from the computational results obtained for these two complexes we can see the following (see Figure 3.3 and Table 3.3). (i) As can be seen from Figure 3.3, when the fullerene C_{60} forms complexes with the $ZnPc(P)_4$ and $ZnPc(S)_4$ species in the gas phase,

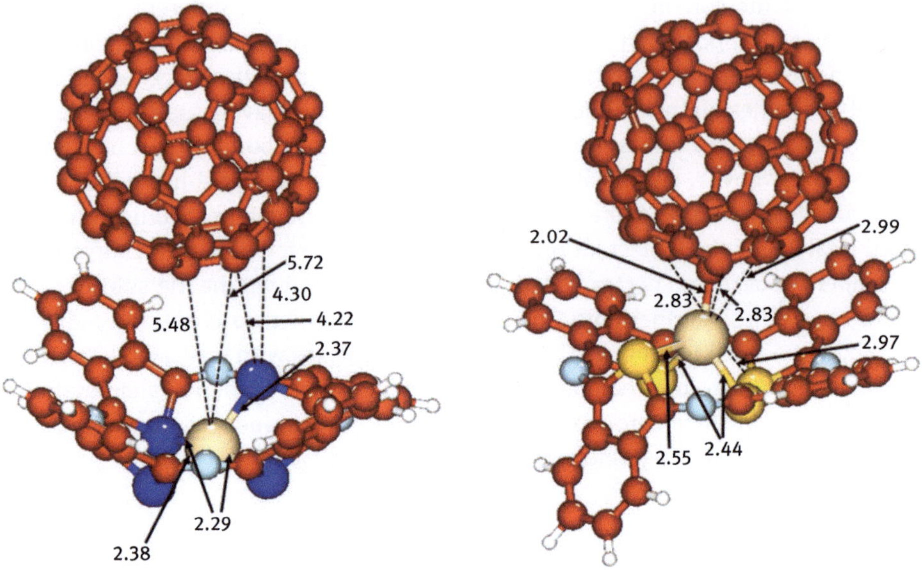

Figure 3.3: Singlet C_{60}-ZnPc(P)$_4$ complex (left) and triplet C_{60}-ZnPc(S)$_4$ complex (right) calculated at the B3LYP/6-31G* level, gas phase, along with the selected bond distances, Å.

Table 3.3: Binding energies, kcal/mol (gas phase), calculated for the C_{60}-ZnPc(X)$_4$ (X = P and S) complexes with the B3LYP/6-31G* // CAM-B3LYP/6-31G* // wB97XD/6-31G* approaches.

Complex	E_{bind}, kcal/mol
C_{60}-ZnPc(P)$_4$	1.2 // 14.2 // 43.6
C_{60}-ZnPc(S)$_4$	13.6 // 26.4 // 52.1

it is located relatively far (4.30–5.72 Å) from the one of the P-centers and from the Zn-center of ZnPc(P)$_4$, whereas with ZnPc(S)$_4$ the fullerene forms relatively short bonds with the Zn-center, varying from ca. 2.0 to ca. 3.0 Å. Also, relatively short distances are observed between the one of the S-centers and the fullerene (not shown in the Figure in order not to overload it): 3.67 – ca. 4.0 Å. (ii) Also, very strong deformations of both the ZnPc(P)$_4$ and ZnPc(S)$_4$ structures are observed, with "opening" of the Pc framework "bowl" and even "bending away" of one of the isoindole units, which is even more pronounced in the case of ZnPc(S)$_4$ species (Figure 3.3). (iii) The Zn-X bonds become noticeably changed compared to the free core-modified Pcs, shortened by ca. 0.05 and elongated by ca. 0.03–0.04 Å in the case of the $ZnPc(P)_4$ species, and strongly elongated by ca. 0.11–0.66 Å in the case of the $ZnPc(S)_4$ species. Also, in the case of the $ZnPc(P)_4$ species, one of the P-centers becomes located closer to the fullerene, like it was observed in the case of formation of the C_{60}-$ZnP(P)_4$ complex [58], and in the case of the $ZnPc(S)_4$ species one Zn-S bond becomes broken, and one of the S-centers becomes located closer to the fullerene as well. (iv) Interesting, in the case of $ZnPc(P)_4$, the singlet structure of the complex formed is lower in energy then the triplet, by 8.9 kcal/mol (see Table S1), whereas for $ZnPc(S)_4$, the triplet structure becomes lower in energy compared to the singlet, by 4.4 kcal/mol (see Table S1). This situation is opposite to the case observed for the complexes formation of the $ZnP(X)_4$ species (X = P and S) with Zn_6S_6, where the singlet complex became much more stable [59]. (v) The calculated binding energy at the B3LYP/6-31G* level for the C_{60}-$ZnPc(P)_4$ complex is quite low, 1.2 kcal/mol, which agrees with the quite long distances fullerene - $ZnPc(P)_4$ (see Figure 3.3), whereas it is noticeably larger, 13.6 kcal/mol, for the C_{60}-$ZnPc(S)_4$ complex which again agrees with the structural features of this complex (Figure 3.3). The binding energies of the complexes optimized using the dispersion-corrected functionals, CAM-B3LYP and wB97XD, are significantly larger, varying from ca. 14 till 52 kcal/mol which corresponds with the shorter distances between the fullerene and $ZnPc(X)_4$ species; thus, for the lowest-lying triplet structure of the C_{60}-$ZnPc(P)_4$ complex calculated using the wB97XD/6-31G* approach (see Table S2) the C_{60}-Zn and C_{60}-P distances becomes noticeably shorter, ca. 3.55–3.6 Å. The binding energies are generally by ca. 8.5–12.4 kcal/mol higher for the C_{60}-$ZnPc(S)_4$ complex (Table 3.3). This might be confirming our suggestion that the $ZnPc(S)_4$ species would react stronger with electron density donors/acceptors.

Due to the fact that no completely core-modified Pcs have been studied experimentally or computationally so far, comparisons of the current study of the non-covalent C_{60}-$ZnPc(X)_4$ (X = P and S) complexes with analogous studies done for the non-covalent complexes of fullerenes with regular Pcs should be performed with certain caution. Thus, the very recent computational study by Bai et al. [88] on the F_nZnPc/C_{60} (n = 0, 4, 8, 16) and Cl_nSubPc/C_{60} (n = 0, 6) complexes showed the binding energies in the complexes varying from 8.83 (for ZnPc-C_{60} complex) till 10.39 ($F_{16}ZnPc$-C_{60} complex) kcal/mol with distances between the fullerene and Pc ranging from 8.26 down to 8.13 Å. The study was done using the wB97XD/6-311G(d,p) approach as implemented in Gaussian 09 with the optimally tuned w, first in the gas phase and solid environment simulated by using a self-consistent reaction field (SCRF) method with the solute electron density model. Another recent theoretical study [72] considered complexation of free-base and 3d transition metal(II) (Mn, Fe, Co, Ni, Cu, Zn) Pcs with endohedral fullerene $Sc_3N@C_{80}$, using the Perdew-Burke-Ernzerhof correlation functional with a long-range dispersion correction by Grimme (PBE-D) as implemented in $DMol^3$. The study provided the binding energies in the complexes investigated varying from ca. 18 till ca. 43 kcal/mol (depending on using the basis set superposition error corrections) with distances between the fullerene and Pc ranging from ca. 2.1 till 2.8 Å. The earlier study by Basiuk and Basiuk [89] on the complexation of free-base and 3d transition metal(II) (Mn, Fe, Co, Ni, Cu, Zn) Pcs with fullerene C_{60} using the PBE GGA functional with a dispersion correction by Grimme gave the binding energies of the complexes varying from ca. 17 up to ca. 37 kcal/mol (depending on using the basis set superposition error corrections) with distances between the fullerene and Pc ranging from ca. 2.1 till ca. 2.9 Å. The 2012 study by Ren et al. [73] was performed on the complex CuPc-C_{60} employing the Ceperley–Alder form of the local density approximation as the exchange-correlation functional, and a local basis set of double-ζ polarized orbitals as implemented in the SIESTA code. Also, vdW-density functionals (vdW-DF) of the Lunqvist–Langreth type for typical bonding configurations were used. The binding energies for the obtained complexes were computed to range from ca. 6.2 till ca. 15 kcal/mol, with distances Pc-C_{60} ranging from the ca. 2.6 till ca 3.0 Å.

Thus, we can conclude that our complexes can potentially exist, they would have quite strongly bound structures, and that the importance of using dispersion-corrected functionals in studying such complexes is quite high. The detailed studies of bonding and molecular orbital pictures of these complexes along with their charge-transfer properties are awaiting to be performed. Also, it will be extremely interesting to see how different Pc metal centers and core-modifying elements along with solvents would influence the complexes formation. This research will be the subject of follow-up investigations, some of which are under way currently.

3.4 Conclusions and perspectives

Pcs and their derivatives have attracted a lot of attention because of their both biological importance and technological applications. Generally speaking, the Pcs and their derivatives have the following characteristic features: (1) a special two-dimensional conjugated 18π-electron structure; (2) high thermal and photostability; (3) they can be modified by a variety of ways; (4) the Pcs coordination ability is very strong. Chemical, optical, coordination, electrochemical, and other properties of Pcs can be easily tuned employing the three major pathways: (i) by replacing the central atom; (ii) by modifying the periphery of Pc ring, and (iii) by changing the *meso*-atoms. One more promising pathway for modifying structures and properties of Pcs and their derivatives can be the so-called *core-modification*, or substitution of the core isoindole nitrogen(s) by other elements. Motivated by the results obtained for core-modified porphyrins and their complexes, we decided to investigate computationally what would occur upon complete core-modification of regular ZnPc with P and S. We performed density functional theory (DFT) studies of the structures, charge, and frontier molecular orbitals of P-core-modified and S-core-modified ZnPcs, $ZnPc(P)_4$ and $ZnPc(S)_4$, respectively, and compared them with the regular ZnPc. We used the following approaches: B3LYP/3-21G*, B3LYP/6-31G*,

CAM-B3LYP/6-31G*, and wB97XD/6-31G* in the gas phase. Also, we studied computationally formation of complexes between the fullerene C_{60} and $ZnPc(P)_4$ and ZnPc $(S)_4$. We used the following computational approaches: B3LYP/3-21G*, B3LYP/6-31G*, CAM-B3LYP/6-31G*, and wB97XD/6-31G* in the gas phase. Results of our study can be summarized as follows.

i. For the two of the three Pc species studied, ZnPc and $ZnPc(S)_4$, singlet structures were found to be the lowest in energy at all the computational approaches employed. However, the situation was found to be different for the $ZnPc(P)_4$ compound: at the B3LYP/6-31G* (B3LYP/3-21G*) the triplet-singlet energy differences were again calculated to be quite noticeable, 10.1 (11.0) kcal/mol. But with the approaches taking the dispersion corrections in the account, the singlet and triplet structures were found to be very close to each other: the singlet-triplet difference calculated with the wB97XD/6-31G* approach was mere 1.5 kcal/mol, and with the CAM-B3LYP/6-31G* approach the triplet structure was calculated to be mere 0.04 kcal/mol lower than the singlet. Thus, for the heavier congener of N, phosphorus, the triplet structure becomes less energetically unfavorable (cf. 10.1 vs. 25.8 kcla/mol for ZnPc) or even energetically comparable with the singlet structure. (Interestingly, for the next heavier congener, As, we found the triplet structure at the B3LYP/6-31G* level to be even closer to the singlet, 5.7 kcal/mol). Thus, extending the carbon framework of the completely core-modified porphyrinic species may cause changes in their ground spin states.
ii. Both $ZnPc(P)_4$ and $ZnPc(S)_4$ show strong bowl-like distortions similar to the results obtained earlier for $ZnP(P)_4$ [54–59] and $ZnP(S)_4$ [59]. From $ZnPc(P)_4$ to $ZnPc(S)_4$, the Zn-center becomes significantly "dented" inside the "bowl cavity" formed by the core-modified Pc moiety which could be ascribed to the bigger S size compared to P and thus making the S4-cavity less suitable to accommodate the Zn-center. The degree of shifting up of the Zn-center in the core-modified Pcs is less compared to the porphyrins core-modified with the same elements (dihedral angles $(X\text{-}X\text{-}X\text{-}Zn)_{av}$ −4.36° for $ZnP(P)_4$ [54, 55] vs. −0.35° for $ZnPc(P)_4$ and −44.83° for $ZnP(S)_4$ [59] vs. −12.17° for $ZnPc(S)_4$). Change of the DFT functional does not affect the value of this upshift significantly. For the triplet structures of $ZnPc(P)_4$ significant alterations in Zn-X bond distances, X-Zn-X angles, and X-X-X-Zn dihedral angles were observed, however, this effect is strongly pronounced only in the P4-cavity: the distances N_1-N_4/N_2-N_3 are essentially the same for both the singlet and triplet $ZnPc(P)_4$ structures. The size of the cavity formed by the porphyrinic ring of the both core-modified Pcs, as judged by the distances N_1-N_4/N_2-N_3, is essentially the same, showing no dependence on the core-modifying element.
iii. There occurs reversal of the charge sign on the X-centers directly connected to the Zn-center for X = P and S, due to much smaller electronegativies of P (2.19) and S (2.58) compared to N (3.04). In general, we can suppose the accumulation of positive charge in the "bottom" part of the core-modified Pc cavity to take

place. From ZnPc to ZnPc(P)$_4$ to ZnPc(S)$_4$, first slight stabilization and then noticeable destabilization of HOMO is observed along with strong stabilization and then even stronger destabilization of LUMO (similar results obtained for MP(P)$_4$ [54, 55]). This, in turn, causes from ZnPc to ZnPc(P)$_4$ to ZnPc(S)$_4$ first strong decrease and then slight increase of the HOMO-LUMO gap. For both the core-modified ZnPc(X)$_4$ the HOMO-LUMO gaps are smaller compared to the ZnPc, by 0.62 (X = P) and 0.29 (X = S) eV. Similar trends in the HOMO-LUMO gap energies are observed with two other DFT functionals. The obtained TDDFT gap values generally support the trends obtained for the HOMO-LUMO gap values. The only difference is for the triplet ZnPc(X)$_4$ TDDFT gap calculated at the wB97XD/6-31G* level: its value is essentially the same as for the wB97XD/6-31G* TDDFT gap for ZnPc. For both ZnPc and ZnPc(P)$_4$ HOMOs are quite similar in shape and in qualitative composition: no visible contributions from Zn, core N/P, and *meso*-N. The LUMOs of these two species, however, have noticeable differences. As for ZnPc(S)$_4$, its HOMO is quite different from those of ZnPc and ZnPc(P)$_4$: some contributions from S- and *meso* N-centers become visible, along with different combinations of orbitals located at the isoindole units. The LUMO of this species has some similarities to the ZnPc(P)$_4$ LUMO showing some contributions from *meso* N-centers, along with slight contributions from the Zn-center and with different combinations of orbitals located at the isoindole units. Such differences in the energetics and compositions of the frontier MOs of ZnPc(P)$_4$ and ZnPc(S)$_4$ compared to ZnPc might lead to different optical/photophysical properties of the core-modified compounds along with their different reactivities. Thus, we can suppose that the destabilized ZnPc(S)$_4$ LUMO would be more available to accept electron density from electron donor (e. g. to the Zn-center orbitals), as well as the destabilized ZnPc(S)$_4$ HOMO would be more available to donate electron density. The accumulation of positive charge in the “bottom” part of the core-modified Pcs “bowl”, along with their strongly pronounced bowl-like deformation, could also lead to their different reactivity and charge transfer properties. More detailed analysis of the molecular orbitals, aromaticity, reactivities, and charge transfer properties of these species is definitely necessary and will be the subject of the follow-up studies. Furthermore, in this study we decided to check the ability of these core-modified Pcs to make non-covalent complexes with the C_{60} fullerene.

iv. For the C_{60}-ZnPc(P)$_4$ complex both with the B3LYP/6-31G* and B3LYP/3-21G* approaches the singlet structure was calculated to be the lowest in energy, by 8.9 and 7.7 kcal/mol, respectively, whereas with the CAM-B3LYP/6-31G* and wB97XD/6-31G* approaches the triplet structure became more energetically favorable, by 5.9 and 4.4 kcal/mol, respectively (see Supporting Information). However, for the C_{60}-ZnPc(S)$_4$ complex the triplet structure is the lowest-lying one with the B3LYP/6-31G*, B3LYP/3-21G*, and CAM-B3LYP/6-31G* approaches, being more stable by 4.4, 1.6, and 11.6 kcal/mol, respectively, but with the wB97XD/6-

31G* approach the singlet becomes lower in energy by mere 0.2 kcal/mol. Thus, from the computational results obtained for these two complexes we can see the following. (i) When the fullerene C_{60} forms complexes with the $ZnPc(P)_4$ and $ZnPc(S)_4$ species in the gas phase, it is located relatively far (4.30–5.72 Å) from the one of the P-centers and from the Zn-center of $ZnPc(P)_4$, whereas with $ZnPc(S)_4$ C_{60} forms relatively short bonds with the Zn-center, varying from ca. 2.0 to ca. 3.0 Å. Also, relatively short distances are observed between the one of the S-centers and the fullerene: 3.67 – ca. 4.0 Å. (ii) Also, very strong deformations of both the $ZnPc(P)_4$ and $ZnPc(S)_4$ structures are observed, with "opening" of the Pc framework "bowl" and even "bending away" of one of the isoindole units. (iii) The Zn-X bonds shortened by ca. 0.05 and elongated by ca. 0.03–0.04 Å in the case of the $ZnPc(P)_4$ species, and strongly elongated by ca. 0.11–0.66 Å in the case of the $ZnPc(S)_4$ species. Also, in the case of the $ZnPc(P)_4$ species, one of the P-centers becomes located closer to the fullerene, like it was observed in the case of formation of the C_{60}-$ZnP(P)_4$ complex [58], and in the case of the $ZnPc(S)_4$ species one Zn-S bond becomes broken, and one of the S-centers becomes located closer to the fullerene as well. (iv) The calculated binding energy at the B3LYP/6-31G* level for the C_{60}-$ZnPc(P)_4$ complex is quite low, 1.2 kcal/mol, which agrees with the quite long distances fullerene - $ZnPc(P)_4$, whereas it is noticeably larger, 13.6 kcal/mol, for the C_{60}-$ZnPc(S)_4$ complex which again agrees with the structural features of this complex. The binding energies of the complexes optimized using the dispersion-corrected functionals, CAM-B3LYP and wB97XD, are significantly larger, varying from ca. 14 till 52 kcal/mol which corresponds with the shorter distances between the fullerene and $ZnPc(X)_4$ species; thus, for the lowest-lying triplet structure of the C_{60}-$ZnPc(P)_4$ complex calculated using the wB97XD/6-31G* approach the C_{60}-Zn and C_{60}-P distances becomes noticeably shorter, ca. 3.55–3.6 Å. The binding energies are generally by ca. 8.5–12.4 kcal/mol higher for the C_{60}-$ZnPc(S)_4$ complex. This might be confirming our suggestion that the $ZnPc(S)_4$ species would react stronger with electron density donors/acceptors.

v. Due to the fact that no completely core-modified Pcs have been studied experimentally or computationally so far, comparisons of the current study of the non-covalent C_{60}-$ZnPc(X)_4$ (X = P and S) complexes with analogous studies done for the non-covalent complexes of fullerenes with regular Pcs should be performed with certain caution. We can conclude that our complexes can potentially exist, they would have quite strongly bound structures, and that the importance of using dispersion-corrected functionals in studying such complexes is quite high.

Based on the obtained computational results, we can formulate the following research perspectives to be addressed:

i. detailed analysis of binding, molecular orbitals, and aromaticity of the core-modified Pcs;

ii. analysis of effects on structures and properties of core-modified Pcs of the metal center type and core-modifying element type;
iii. detailed investigation of formation of complexes of core-modified Pcs with various NPs, fullerenes, and smaller molecules, understanding the bonding and stability of such complexes and ways of tuning their structures and properties.

Funding: The computational resources of the Centro Nacional de Processamento de Alto Desempenho – UFC and of the computational center in ITA are highly appreciated.

Supporting Information

Energies of the $ZnPc(X)_4$ species (X = N, P, S) and their frontier MOs, calculated at B3LYP/6-31G*, B3LYP/3-21G*, CAM-B3LYP/6-31G*, and wB97XD/6-31G* levels of theory [gas phase]. Energies of the C_{60} and C_{60}-$ZnPc(X)_4$ (X = P, S) and their frontier MOs calculated at the B3LYP/6-31G*, B3LYP/3-21G*, CAM-B3LYP/6-31G*, and wB97XD/6-31G* levels of theory [gas phase]. These data can be received from the author by request.

References

[1] Leznoff CC, Lever AB. Phthalocyanines properties and applications, vol. 1. New York: VCH Publisher, 1989. vol. 2, 1993, vol. 3, 1993.
[2] Kadish KM, Smith KM, Guilard R. The porphyrin handbook: applications of phthalocyanines Vol. 19. San Diego: Academic Press, 2003.
[3] Jiang J. Functional phthalocyanine molecular materials. Germany: Springer-VBH, 2010.
[4] Gutzler R, Perepichka DF. π-Electron conjugation in two dimensions. J Am Chem Soc. 2013;135:16585–94.
[5] Liao MS, Kar T, Gorun SM, Scheiner S. Effects of peripheral substituents and axial ligands on the electronic structure and properties of iron phthalocyanine. Inorg Chem. 2004;43:7151–61.
[6] Konarev DV, Kuzmin AV, Nakano Y, Faraonov MA, Khasanov SS, Otsuka A, et al. Coordination complexes of transition metals (M = Mo, Fe, Rh, and Ru) with Tin(II) phthalocyanine in neutral, monoanionic, and dianionic states. Inorg Chem. 2016;55:1390–402.
[7] Chen J, Zhu C, Xu Y, Zhang P, Liang T. Advances in phthalocyanine compounds and their photochemical and electrochemical properties. Curr Org Chem. 2018;22:485–504. DOI: 10.2174/1385272821666171002122055.
[8] Gerasymchuk YS, Tomachynski LN, Tretyakowa LN, Hanuza J, Legendziewicz J. Axially substituted ytterbium(III) monophthalocyanine—synthesis and their spectral properties in solid state, solution and in monolithic silica blocks. J Photochem Photobiol A. 2010;214:128–34.
[9] Godlewska P, Gerasymchuk YS, Tomachynski LN, Legendziewicz J, Hanuza J. Molecular structure of phthalocyaninato lanthanide LnPc(OAc) complexes derived from the FTIR and FT Raman studies. J Struct Chem. 2010;21:461–7.

[10] Günsel A, Bilgiçli AT, Kandaz M, Orman EB, Özkaya AR. Ag(I) and Pd(II) sensing, H- or J-aggregation and redox properties of metal-free, manganase(III) and gallium(III) phthalocyanines. Dyes Pigm. 2014;102:169–79. DOI: 10.1016/j.dyepig.2013.09.035.
[11] Osifeko OL, Durmuş M, Nyokong T. Physicochemical and photodynamic antimicrobial chemotherapy studies of mono- and tetra-pyridyloxy substituted indium(III) phthalocyanines. J Photochem Photobiol A. 2015;301:47–54. DOI: 10.1016/j.jphotochem.2014.12.011.
[12] San SE, Okutan M, Nyokong T, Durmus M, Ozturk B. Temperature activated ionic conductivity in gallium and indium phthalocyanines. Polyhedron. 2011;30:1023–6. DOI: 10.1016/j.poly.2010.12.047.
[13] Thimiopoulos A, Vogiatzi A, Simandiras ED, Mousdis GA, Psaroudakis N. Synthesis, characterization and DFT analysis of new phthalocyanine complexes containing sulfur rich substituents. Inorg Chim Acta. 2019;488:170–81. DOI: 10.1016/j.ica.2019.01.010.
[14] Dai G, Chen L, Xie F. Structures and electronic properties of halogenated Au(III) phthalocyanine AuPcX (X = Cl, Br): a density functional theoretical study. Comp Mater Sci. 2018;152:262–7. DOI: 10.1016/j.commatsci.2018.06.007.
[15] Chen X, Wang C, Chen Y, Qi D, Jiang J. Vibrational spectra of alkylamino substituted phthalocyanine compounds: density functional theory calculations. J Porph Phthalocyanines. 2018;22:1–6. DOI: 10.1142/S1088424618500591.
[16] Güzel E, Günsel A, Tüzün B, Atmaca GY, Bilgiçli AT, Erdoğmuş A, et al. Synthesis of tetra-substituted metallophthalocyanines: spectral, structural, computational studies and investigation of their photophysical and photochemical properties. Polyhedron. 2019;158:316–24. DOI: 10.1016/j.poly.2018.10.072.
[17] Kim C, Kumar RS, Mergu N, Jun K, Son YA. Synthesis, optical, electrochemical and theoretical studies of new multicyclic substituted phthalocyanines. J Nanosci Nanotechnol. 2018;18:3192–205. DOI: 10.1166/jnn.2018.14855.
[18] Lukyanets EA, Nemykin VN. The key role of peripheral substituents in the chemistry of phthalocyanines and their analogs. J Porph Phthalocyanines. 2010;14:1–40.
[19] Nemykin VN, Lukyanets EA. Synthesis of substituted phthalocyanines. ARKIVOC. 2010;I:136–208.
[20] Çimen Y, Ermiş E, Dumludağ F, Özkaya AR, Salih B, Bekaroğlu Ö. Synthesis, characterization, electrochemistry and VOC sensing properties of novel ball-type dinuclear metallophthalocyanines. Sens Actuators B Chem. 2014;202:1137–47. DOI: 10.1016/j.snb.2014.06.066.
[21] Yildiz B, Güzel E, Menges N, Şişman İ, Şener MK. Pyrazole-3-carboxylic acid as a new anchoring group for phthalocyanine-sensitized solar cells. Sol Energy. 2018;174:527–36. DOI: 10.1016/j.solener.2018.09.039.
[22] Harrath K, Talib SH, Boughdiri S. Theoretical design of metal-phthalocyanine dye-sensitized solar cells with improved efficiency. J Mol Modeling. 2018;24:279.
[23] Ikeuchi T, Nomoto H, Masaki N, Griffith MJ, Mori S, Kimura M. Molecular engineering of zinc phthalocyanine sensitizers for efficient dye-sensitized solar cells. Chem Commun. 2014;50:1941–3. DOI: org/10.1007/s00894-018-3821-6.
[24] Bottari G, de la Torre G, Guldi DM, Torres T. Covalent and noncovalent phthalocyaninecarbon nanostructure systems: synthesis, photoinduced electron transfer, and application to molecular photovoltaics. Chem Rev. 2010;110:6768–816.
[25] Tietze ML, Tress W, Pfützner S, Schünemann C, Burtone L, Riede M, et al. Correlation of open-circuit voltage and energy levels in zinc-phthalocyanine: C_{60} bulk heterojunction solar cells with varied mixing ratio. Phys Rev B. 2013;88:085119.
[26] Ray A, Santhosh K, Bhattacharya S. Spectroscopic and structural insights on molecular assembly consisting high potential zinc phthalocyanine photosensitizer attached to PyC_{60} through non-covalent interaction. Spectrochim Acta A. 2015;135:386–97.

[27] Günsel A, Güzel E, Bilgiçli AT, Şişman İ, Yarasir MN. Synthesis of nonperipheral thioanisole-substituted phthalocyanines: photophysical, electrochemical, photovoltaic, and sensing properties. J Photochem Photobio A: Chem. 2017;348:57–67.

[28] de la Torre G, Bottari G, Torres T. Phthalocyanines and subphthalocyanines: perfect partners for fullerenes and carbon nanotubes in molecular photovoltaics. Adv Energy Mater. 2017;7:1601700. DOI: 10.1002/aenm.201601700.

[29] Gao F, Yang CL, Wang MS, Ma XG, Liu WW. Theoretical studies on the possible sensitizers of DSSC: nanocomposites of graphene quantum dot hybrid phthalocyanine/ tetrabenzoporphyrin/ tetrabenzotriazaporphyrins/cis-tetrabenzodiazaporphyrins/ tetrabenzomonoazaporphyrins and their Cu-metallated macrocycles. Spectrochim Acta A: Mol Biomol Spectr. 2018;195:176–83. DOI: 10.1016/j.saa.2018.01.065.

[30] Sorokin AB. Phthalocyanine Metal Complexes in Catalysis. Chem Rev. 2013;113:8152–91. DOI: 10.1021/cr4000072.

[31] Kamiloglu AA, Acar İ, Biyiklioglu Z, Saka ET. Peripherally tetra-{2-(2,3,5,6-tetrafluorophenoxy) ethoxy} substituted cobalt(II), iron(II) metallophthalocyanines: synthesis and their electrochemical, catalytic activity studies. J Organomet Chem. 2017;828:59–67.

[32] Souza JS, Pinheiro MV, Krambrock K, Alves WA. Dye degradation mechanisms using nitrogen doped and copper(II) phthalocyanine tetracarboxylate sensitized titanate and TiO_2 nanotubes. J Phys Chem C. 2016;120:11561–71.

[33] Khoza P, Nyokong T. Photocatalytic behaviour of zinc tetraamino phthalocyanine-silver nanoparticles immobilized on chitosan beads. J Mol Catal A Chem. 2015;399:25–32.

[34] Alsudairi A, Li J, Ramaswamy N, Mukerjee S, Abraham KM, Jia Q. Resolving the iron phthalocyanine redox transitions for ORR catalysis in aqueous media. J Phys Chem Lett. 2017;8:2881–6.

[35] Sytnyk M, Głowacki ED, Yakunin S, Voss G, Schołfberger W, Kriegner D, et al. Hydrogen-bonded organic semiconductor micro- and nanocrystals: from colloidal syntheses to (Opto-) electronic devices. J Am Chem Soc. 2014;136:16522–32. DOI: 10.1021/ja5073965.

[36] Ostroverkhova O. Organic optoelectronic materials: mechanisms and applications. Chem Rev. 2016;116:13279–13412. DOI: 10.1021/acs.chemrev.6b00127.

[37] Basova TV, Parkhomenko RG, Polyakov M, Gürek AG, Atilla D, Yuksel F, et al. Effect of dispersion of gold nanoparticles on the properties and alignment of liquid crystalline copper phthalocyanine films. Dye Pigment. 2016;125:266–73. DOI: 10.1016/j.dyepig.2015.10.005.

[38] Harish TS, Viswanath P. Annealing assisted structural and surface morphological changes in Langmuir–blodgett films of nickel octabutoxy phthalocyanine. Thin Solid Films. 2016;598:170–6. DOI: 10.1016/j.tsf.2015.11.065.

[39] Melville OA, Lessard BH, Bender TP. Phthalocyanine-based organic thin-film transistors: A review of recent advances. ACS Appl Mater Interf. 2015;7:13105–18.

[40] Lu H, Kobayashi N. Optically active porphyrin and phthalocyanine systems. Chem Rev. 2016;116:6184–6261. DOI: 10.1021/acs.chemrev.5b00588.

[41] Haldar S, Bhandary S, Vovusha H, Sanyal B. Comparative study of electronic and magnetic properties of iron and cobalt phthalocyanine molecules physisorbed on two-dimensional MoS_2 and graphene. Phys Rev B. 2018;98:085440.

[42] Jurow M, Schuckman AE, Batteas JD, Drain CM. Porphyrins as molecular electronic components of functional devices. Coord Chem Rev. 2010;254:2297–310.

[43] Tykarska E, Wierzchowski M, Teubert A, Wyszomirska FA, Wyszko E, Gdaniec M, et al. Phthalocyanine derivatives possessing 2-(morpholin-4-yl)ethoxy groups as potential agents for photodynamic therapy. J Med Chem. 2015;58:2240–55.

[44] Şen P, Atmaca GY, Erdogmus A, Kanmazalp SD, Dege N, Yildiz SZ. Peripherally tetra-benzimidazole units-substituted zinc(II) phthalocyanines: synthesis, characterization and

investigation of photophysical and photochemical properties. J Luminescence. 2018;194:123–30.

[45] Günsel A, Beylik S, Bilgiçli AT, Atmaca GY, Erdogmus A, Yarasir MN. Peripherally and non-peripherally tetra-HBME (4-hydroxybenzyl methyl ether) substituted metal-free and zinc(II) phthalocyanines: synthesis, characterization, and investigation of photophysical and photochemical properties. Inorg Chim Acta. 2018;477:199–205.

[46] Günsel A, Bilgiçli AT, Kırbaç E, Güney S, Kandaz M. Water soluble quarternizable gallium and indium phthalocyanines bearing quinoline 5-sulfonic acid: synthesis, aggregation, photophysical and electrochemical studies. J Photochem Photobio A: Chem. 2015;310:155–64.

[47] Kucinska M, Mrugalska SP, Szczolko W, Sobotta L, Sciepura M, Tykarska E, et al. Phthalocyanine derivatives possessing 2-(morpholin-4-yl)ethoxy groups as potential agents for photodynamic therapy. J Med Chem. 2015;58:2240–55.

[48] Wang K, Qi D, Li Y, Wang T, Liu H, Jiang J. Tetrapyrrole macrocycle based conjugated two-dimensional mesoporous polymers and covalent organic frameworks: from synthesis to material applications. Coord Chem Rev. 2019;378:188–206. DOI: org/10.1016/j.ccr.2017.08.023.

[49] Latos-Grażyński L. Core modified heteroanalogue of porphyrins and metalloporphyrins. In: Kadish KM, Smith KM, Guilard R, editors. The porphyrin handbook. New York: Academic Press, 2000:361–416.

[50] Gupta I, Ravikanth M. Recent developments in heteroporphyrins and their analogues. Coord Chem Re. 2006;250:468–518.

[51] Szyszko B, Latos-Grażyński L. Core chemistry and skeletal rearrangements of porphyrinoids and metalloporphyrinoids. Chem Soc Rev. 2015;44:3588–3616.

[52] Chatterjee T, Shetti VS, Sharma R, Ravikanth M. Heteroatom-containing porphyrin analogues. Chem Rev. 2017;117:3254–3328.

[53] Kuznetsov AE. Design of novel classes of building blocks for nanotechnology: core-modified metalloporphyrins and their derivatives. In: Akitsu T, editor(s). Descriptive inorganic chemistry researches of metal compounds. Croatia: InTechOpen, 2017:135–52. 978-953-51-3398-8.

[54] Barbee J, Kuznetsov AE. Revealing substituent effects on the electronic structure and planarity of Ni-porphyrins. Comp Theor Chem. 2012;981:73–85.

[55] Kuznetsov AE. Metalloporphyrins with all the pyrrole nitrogens replaced with phosphorus atoms, $MP(P)_4$ (M = Sc, Ti, Fe, Ni, Cu, Zn). Chem Phys. 2015;447:36–45.

[56] Kuznetsov AE. How the change of the ligand from L = porphine, P^{2-}, to L = P_4-substituted porphine, $P(P)_4{}^{2-}$, affects the electronic properties and the M-L binding energies for the first-row transition metals M = Sc-Zn: comparative study. Chem Phys. 2016;469–470:38–48.

[57] Kuznetsov AE. Computational design of $ZnP(P)_4$ stacks: three modes of binding. J Theor Comput Chem. 2016;15:1650043.

[58] Kuznetsov AE. Can $MP(P)_4$ compounds form complexes with C_{60}? J Appl Sol Chem Model. 2017;6:91–7.

[59] Kuznetsov AE. Complexes between core-modified porphyrins $ZnP(X)_4$ (X = P and S) and small semiconductor nanoparticle Zn_6S_6: are they possible? DOI: 10.1515/psr-2017-0187.

[60] Frisch MJ, Trucks GW, Schlegel HB, Scuseria GE, Robb MA, Cheeseman JR, et al. Gaussian 09: ES64L-G09RevD.01 24-Apr-2013. Wallingford, CT: Gaussian, Inc., 2013.

[61] Parr RG, Yang W. Density-functional theory of atoms and molecules. Oxford: Oxford University Press, 1989.

[62] Ditchfield R, Hehre WJ, Pople JA. Self-consistent molecular-orbital methods. IX. An extended Gaussian-type basis for molecular-orbital studies of organic molecules. J Chem Phys. 1971;54:724–8.

[63] Hehre WJ, Ditchfield R, Pople JA. Self-consistent molecular orbital methods. XII. Further extensions of Gaussian-type basis sets for use in molecular orbital studies of organic molecules. J Chem Phys. 1972;56:2257–61.

[64] Hariharan PC, Pople JA. Accuracy of AH_n equilibrium geometries by single determinant molecular orbital theory. Mol Phys. 1974;27:209–14.

[65] Gordon MS. The isomers of silacyclopropane. Chem Phys Lett. 1980;76:163–8.

[66] Hariharan PC, Pople JA. The influence of polarization functions on molecular orbital hydrogenation energies. Theo Chim Acta. 1973;28:213–22.

[67] Kozlowski PM, Bingham JR, Jarzecki AA. Theoretical analysis of core size effects in metalloporphyrins. J Phys Chem A. 2008;112:12781–8.

[68] Myradalyyev S, Limpanuparb T, Wang X, Hirao H. Comparative computational analysis of binding energies between several divalent first-row transition metals (Cr^{2+}, Mn^{2+}, Fe^{2+}, Co^{2+}, Ni^{2+}, and Cu^{2+}) and ligands (porphine, corrin, and TMC). Polyhedron. 2013;52:96–101.

[69] Kaura P, Sachdeva R, Singh R, Soleimanioun N, Singh S, Saini GS. Effect of asymmetrical peripheral substitution of sulfonic acid group on the geometric and electronic structures and vibrations of copper phthalocyanine studied by Computational and Experimental techniques. J Mol Struc. doi:10.1016/j.molstruc.2018.07.082.

[70] Tverdova NV, Girichev GV, Krasnov AV, Pimenov OA, Koifman OI. The molecular structure, bonding, and energetics of oxovanadium phthalocyanine: an experimental and computational study. Struct Chem. 2013;24:883–90.

[71] Albakoura M, Tunça G, Akyola B, Kostakoğlu ST, Berber S, Bekaroglu O, et al. Synthesis, characterization, photophysicochemical properties and theoretical study of novel zinc phthalocyanine containing four tetrathia macrocycles. J Porph Phthalocyanines. 2018;22:77–87.

[72] Basiuk VA, Tahuilan-Anguiano DE. Complexation of free-base and 3d transition metal(II) phthalocyanines with endohedral fullerene $Sc_3N@C_{80}$. Chem Phys Lett. 2019;722:146–52. DOI: org/10.1016/j.cplett.2019.03.019.

[73] Ren J, Meng S, Kaxiras E. Theoretical investigation of the C_{60}/copper phthalocyanine organic photovoltaic heterojunction. Nano Res. 2012;5:248–57.

[74] Qi D, Zhang L, Wan L, Zhao L, Jiang J. Design of a universal reversible bidirectional current switch based on the fullerene–phthalocyanine supramolecular system. J Phys Chem A. 2012;116:6785–91.

[75] Das SK, Mahler A, Wilson AK, D'Souza F. High-potential perfluorinated phthalocyanine–fullerene dyads for generation of high-energy charge-separated states: formation and photoinduced electron-transfer studies. Chem Phys Chem. 2014;15:2462–72.

[76] Sai N, Gearba R, Dolocan A, Tritsch JR, Chan WL, Chelikowsky JR, et al. Understanding the interface dipole of copper phthalocyanine (CuPc)/C_{60}: theory and experiment. J Phys Chem Lett. 2012;3:2173–7.

[77] Guo JH, Zhang H, Miyamoto Y. New Li-doped fullerene-intercalated phthalocyanine covalent organic frameworks designed for hydrogen storage. Phys Chem Chem Phys. 2013;15:8199–207.

[78] Ray A, Goswami D, Chattopadhyay S, Bhattacharya S. Photophysical and theoretical investigations on fullerene/phthalocyanine supramolecular complexes. J Phys Chem A. 2008;112:11627–40.

[79] Shalabi AS, Aal SA, Assem MM, Soliman KA. Metallophthalocyanine and metallophthalocyanine–fullerene complexes as potential dye sensitizers for solar cells DFT and TD-DFT calculations. Org Electronics. 2012;13:2063–74.

[80] Ray A, Santhosh K, Bhattacharya S. Absorption spectrophotometric, fluorescence, transient absorption and quantum chemical investigations on fullerene/phthalocyanine supramolecular complexes. Spectrochim Acta A. 2011;78:1364–75.

[81] Yanai T, Tew D, Handy N. A new hybrid exchange-correlation functional using the Coulomb-attenuating method (CAM-B3LYP). Chem Phys Lett. 2004;393:51–7. DOI: 10.1016/j.cplett.2004.06.011.
[82] Chai JD, Head-Gordon M. Long-range corrected hybrid density functionals with damped atom-atom dispersion corrections. Phys Chem Chem Phys. 2008;10:6615–20. DOI: 10.1039/B810189B.
[83] Binkley JS, People JA, Hehre WJ. Self-consistent molecular orbital methods. 21. Small split-valence basis sets for first-row elements. J Am Chem Soc. 1980;102:939–47.
[84] Dobbs KD, Hehre WJ. Molecular-orbital theory of the properties of inorganic and organometallic compounds. 6. Extended basis-sets for 2nd-row transition-metals. J Comp Chem. 1987;8:880–93.
[85] Reed AE, Curtiss LA, Weinhold F. Intermolecular interactions from a natural bond orbital, donor-acceptor viewpoint. Chem Rev. 1988;88:899–926.
[86] Reed AE, Weinstock RB, Weinhold F. Natural-population analysis. J Chem Phys. 1985;83:735–46.
[87] Schaftenaar G, Noordik JH. Molden: a pre- and post-processing program for molecular and electronic structures. J Comput-Aided Mol Design. 2000;14:123–34.
[88] Bai RR, Zhang CR, Wu YZ, Shen YL, Liu ZJ, Chen HS. Donor halogenation effects on electronic structures and electron process rates of donor/c_{60} heterojunction interface: a theoretical study on F_nZnPc (n = 0, 4, 8, 16) and Cl_nSubPc (n = 0, 6). J Phys Chem A. 2019;123:4034–4047. DOI: 10.1021/acs.jpca.9b01937.
[89] Basiuk VA, Basiuk EV. Complexation of free-base and 3d transition metal(II) phthalocyanines with fullerene C_{60}: a dispersion-corrected DFT study. Nanotubes Carbon Nanostruc. 2017;25:410–16. DOI: 10.1080/1536383x.2017.1325363.

A. Annam Renita and V. Sivasubramanian

4 Application of computational chemistry for adsorption studies on metal–organic frameworks used for carbon capture

Abstract: Computational chemistry is invaluable in calculating macroscopic and microscopic details of systems application in chemical industries which are involved in carbon capture through precombustion, post-combustion and oxy combustion technologies. This review discusses the role of computational chemistry for adsorption studies of metal–organic frameworks (MOFs) which can be utilized for carbon capture. Principles of quantum mechanics–molecular mechanics are used to devise the electrostatic charges and isotherm parameters on the MOFs. MOFs for carbon capture which can be compatible and which can withstand the severity in chemical industries can be effectively studied using grand canonical Monte Carlo simulation by selecting appropriate force fields. Since flue gases contain a host of other gases in addition to oxides of carbon, capture by MOFs has to be carefully modelled and the software useful for this study are mentioned in this review. The simulated adsorption isotherms should be compared with experimental adsorption isotherms to validate the study. The adsorption model for carbon dioxide adsorption on MOFs is generally reported to be type I reversible isotherm and the kinetics is in good agreement with pseudo-second-order kinetics.

Keywords: carbon dioxide, adsorption, MOF, GCMC, software

This article has previously been published in the journal Physical Sciences Reviews. Please cite as: Renita, A. A., Sivasubramanian, V. Application of computational chemistry for adsorption studies on metal–organic frameworks used for carbon capture *Physical Sciences Reviews* [Online] 2020, 5 DOI: 10.1515/psr-2018-0179.

https://doi.org/10.1515/9783110631623-004

Graphical Abstract:

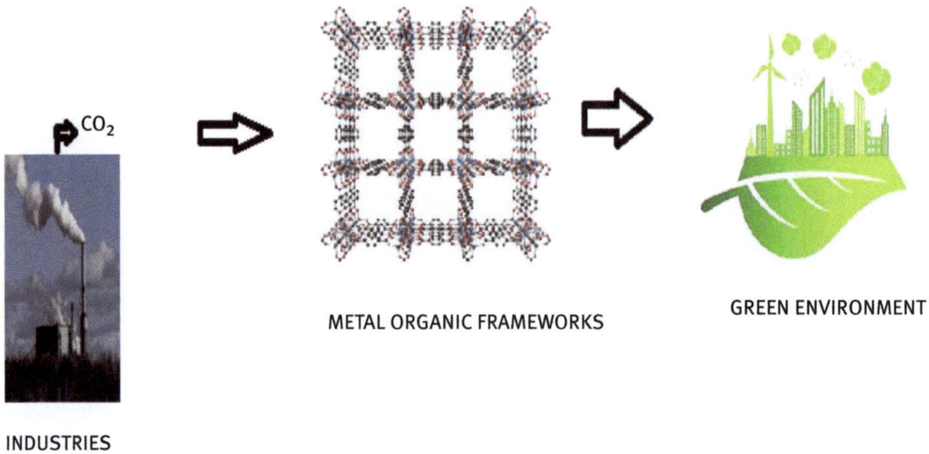

4.1 Introduction

Reduction of greenhouse gases has become the prime focus of environmentalists worldwide. Carbon dioxide, a greenhouse gas, plays a major role as it influences the biogeochemical cycles of the environment. Many countries who are members of the Mission Innovation have pledged to reduce their carbon footprint by 2020 in United Nations Framework Convention on Climate Change Paris Summit. The Panel on Climate Change estimates that emissions of CO_2 can be reduced by 80–90% with carbon capture and storage technology for a conventional power plant [1].

Carbon dioxide can be captured and stored or can be captured and utilized, thereby balancing the carbon cycle. In both the processes, effective carbon capture is the pressing issue since it is highly challenging. The key factor which underlies significant advancements lies in improved materials that perform the separations which is achieved by solvent absorption, adsorption, cryogenic and membrane separation. Among the four, adsorption method is corrosion free and is a low energy input process. Many adsorbents have been reported so far for carbon capture. The most effective and versatile adsorbents are metal organic frameworks (MOFs) because it can be tailor made to adapt to the differing compositions of CO_2 gas in the exhaust. MOFs are nanoporous, crystalline materials, self-assembled from inorganic metal "nodes" and organic "linkers" [2, 3]. High pore volume, tailorable pore sizes and large surface areas make MOFs the most suitable choice for gas separations [4]. Due to the unlimited probability of building blocks and linkers, the research on new MOFs is challenging and a tool to model them by computational method is highly recommended. Computational chemistry techniques can serve as a powerful tool to complement or act as an alternative to experiment, to help design and analyze the performance of MOFs more rapidly than with experiments alone [5, 6]. This review paper is just a drop in the ocean of computational chemistry methods and principles available for adsorption studies of MOFs used for carbon capture as the systemic review is beyond the scope attributed to vast expanse of the field.

4.2 Conventional carbon capture methods

The emission of carbon dioxide can be controlled in three stages – precombustion, post-combustion and oxy combustion. Various industries will have either modification in these systems or an adoption of one or more than one of the technologies mentioned below.

4.2.1 Precombustion system

Fuels when combusted with air in a gasifier release a combination of gases comprising carbon monoxide (CO), carbon dioxide (CO_2), hydrogen (H_2), methane (CH_4), etc. These exhaust gases are further sent into a reactor where water gas shift reaction takes place with the help of steam to give synthesis gas (CO, H_2, CO_2). Synthesis gas is sent to a turbine system to generate electricity and this concept is used in power plants. The remaining CO_2 is captured via absorption or adsorption or membrane separation methods and is further stored in geological reservoirs.

4.2.2 Post-combustion system

Exhaust gases from boilers or from various processes contain CO_2 in the flue gas which is generally flared through tall stacks. Most of the manufacturing industries follow this procedure. CO_2 from the flue gas can be separated by either of the four methods – absorption, adsorption, membrane separation and cryogenic separation. The CO_2 thus captured is stored in underground reservoirs.

4.2.3 Oxy combustion system

This process utilizes pure oxygen for combustion by separation of other gases like nitrogen, hydrogen, etc., from air. When pure oxygen and fuel undergo combustion, they produce relatively clean CO_2, water (H_2O) and small amounts of oxides of sulfur and nitrogen (SO_x, NO_x), mercury, particulates along with enormous amount of heat energy of the magnitude of 2,000–3,000 °F. CO_2 so liberated can be compressed and sequestered [7]. High heat energy produced in the process is utilized in glass and metallurgical industries. Major setbacks are the cost and energy consumption incurred in separation of oxygen from air. The carbon dioxide thus captured is sequestrated and stored in subsurface geologic formations, at 100–150 bar in depleted oil and gas fields at around 800–1,000 m below the surface [8].

The schematic representation is given below for the above three technologies in Figure 4.1–Figure 4.3.

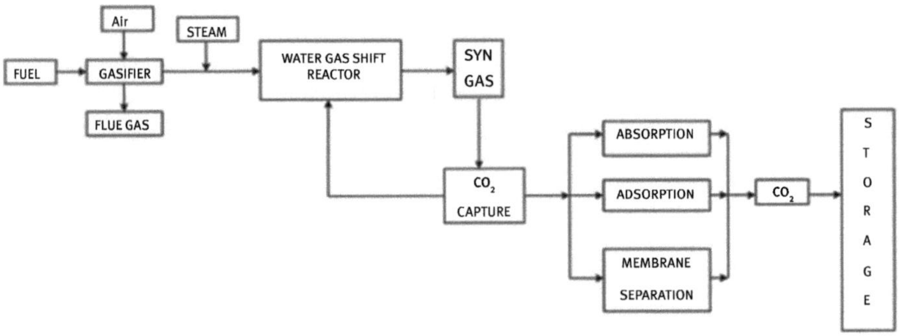

Figure 4.1: Block diagram for precombustion of carbon technology.

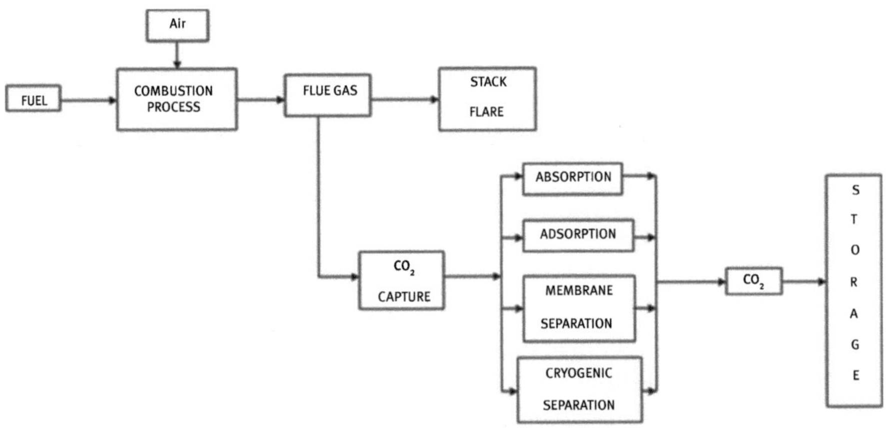

Figure 4.2: Block diagram for post-combustion of carbon technology.

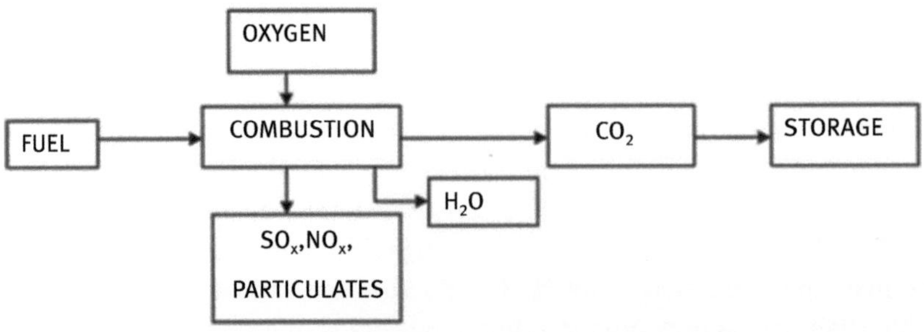

Figure 4.3: Block diagram for oxy combustion of carbon technology.

4.3 Adsorption

The methods available for carbon capture are absorption, adsorption, cryogenic separation and membrane separation. The conventional method which is commonly used in industries is amine absorption and it has energy penalties associated with it which makes the process uneconomical [9]. Membrane processes have high capital costs and life of membranes is limited. Cryogenic separations involve high capital and maintenance costs. The pressure swing adsorption (PSA) system is known to be one of the most efficient and economical processes to recover carbon dioxide in flue streams from power plants and incinerators and the adsorbents for CO_2 play a crucial role in effectively capturing gas and for cost reduction [10]. Moreover, they can be regenerated and they utilize less energy compared to its counterparts in regeneration. The commonly used adsorbents for carbon capture are carbon-based sorbents such as activated carbon and carbon molecular sieves, zeolites, chemically modified mesoporous materials, metal oxides, MOFs and hydrotalcite-like compounds, amongst others [11, 12].

4.3.1 Mechanism

Adsorption is basically a surface phenomena and is classified as physical adsorption and chemical adsorption. Physical adsorption is characterized by weak van der Waals forces and low heat of adsorption, and chemical adsorption is characterized by strong covalent bonds and high heat of adsorption. CO_2 adsorption on MOFs has been mostly reported to be physisorption [13, 14] which can be desorbed to recover pure CO_2 which can be used for fuel production, dry ice manufacture and a host of speciality chemicals.

4.3.2 Isotherms and kinetics

The extent of adsorption is usually studied by adsorption isotherms. Its physicochemical parameters together with the underlying thermodynamic assumptions provide an insight into the adsorption mechanism, surface properties as well as the degree of affinity of the adsorbents [15]. Over the years, a wide variety of equilibrium isotherm models namely Henry, Langmuir, Freundlich, Fowler–Guggenheim Model, Jovanovic, Elovich, Kiselev, Jossens, Fritz–Schlunder, Baudu, Weber–Van Vliet, Marczewski–Jaroniec, Fritz and Schlunder, Halsey, Harkin–Jura, Langmuir–Freundlich, Brunauer–Emmett–Teller, Redlich–Peterson, Dubinin–Radushkevich, Temkin, Toth, Koble–Corrigan, Sips, Khan, Hill, Flory–Huggins and Radke–Prausnitz isotherms have been formulated based on the number of parameters involved [16, 17]. The adsorption model for CO_2 adsorption on MOFs is generally reported to be type I reversible (Langmuir) isotherm [11, 18]. Kinetic studies explain the rate of adsorption and give valuable data on the order of the adsorption and equilibrium constant. CO_2 adsorption on MOFs has been reported to be in good agreement with pseudo-second-order kinetics [19, 20].

4.4 MOFs in carbon capture

MOFs find application in molecular sensing [21, 22], gas storage and separation [23–26] heterogeneous catalysis [22, 27] and drug delivery [28]. The most viable and promising materials for all types of carbon capture processes are MOFs. Many researchers have reported that MOFs have higher CO_2 selectivities and higher CO_2 working capacities than zeolites and carbon-based adsorbents [29]. MOFs are composed of elementary or secondary building units assembled together into three-dimensional networks [30, 31]. MOFs correspond to hybrid crystalline solids with periodic nanoporous architecture in which inorganic metal-ion-based nodes are bridged by polytopic organic ligands through coordination bonds [32]. The high surface area-to-weight ratio of MOFs enhances CO_2 capture at moderate pressures and room temperature compared to other adsorbents [12]. Molecules of carbon dioxide are nonpolar and have an intrinsic quadrupole moment, thus making MOFs highly desirable for capturing CO_2 molecules in its internal surface area [33]. The MOFs reported to have good carbon uptake at room temperature are [Mg-MOF-74] [34], [MOF-1774] [35], [Cu-TDPAT], [SIFSIX-Cu-2-i] [36], [H6TDPAT], [2,4,6-tris(3,5-dicarboxylphenylamino)-1,3,5-triazine] [37], [$Zn_4O(btb)_2$], [$Zn_4O(bdc)_3$] [24], [$Cu_3(btc)_2$], [Mg(tcpbda)] [38], [Al(OH)(bdc)] [39], UiO-66, NU-1000 [40], [$Ni_2(bdc)_2(dabco)$], [In(OH)(NH_2bdc)] [In(OH)(bdc)] [41] and MIL-53 [11]. MOFs with high surface areas and pore diameters of greater than 15 Å have generally exhibited high uptake of CO_2. The framework [$Zn_4O(btb)_2$] which exhibits the highest capacity for CO_2 capture has a surface area of 4,500 $m^2\,g^{-1}$ and adsorbs up to 33.5 mmol g^{-1} CO_2 at 32 bar [25].

4.5 Role of computational chemistry in carbon capture

Computational chemistry has been useful in accurately predicting the adsorption-based CO_2 separation potentials of MOFs [21, 22]. Computational chemistry can give molecular level insights which will facilitate the researcher to achieve the desired result by modifications of pore volume, surface area and functionality of the materials used for carbon capture. Moreover, generation of millions of hypothetical MOF structures is possible which will help in high-throughput screening to identify potential synthesis targets [42]. The computational tool which aptly helps in studying the complex mechanism of CO_2 adsorption onto a MOF is a combined quantum mechanical–molecular mechanics (MM) approach. The major advantage of computational modelling is that the key portions like active sites for adsorption can also be studied using ab-initio approach [43] or semiempirical approach [44]. Active sites of adsorbent are specific points on the adsorbent which forms strong bonds with the adsorbate which is generally fluid.

4.5.1 Force fields

MOF versatility cannot be restricted to conventional force field since their structure can be transformed to adapt to their field application. In most of the computational studies regarding MOFs used for carbon capture, the framework atoms are assumed to be fixed at their crystallographic positions. Though this assumption does not affect accuracy, it makes the simulation simpler by neglecting intra-framework interactions. The energy calculations from quantum mechanics (QM) can be used to describe force fields for specific class of MOFs using Morse or other potentials for thermodynamic interactions [45, 46]. They are polarizable, nonpolarizable, all-atom, coarse-grained, diagonal and cross term type force fields [47]. When generic force fields are used, testing with experiments is mandatory to check the applicability of the force field for particular systems of concern. The atomic level interactions of adsorbate and the bonds established between adsorbate gas molecules and the adsorbent MOFs play a major role in deciding the force fields often used for MOFs. A library of well-established force fields for adsorbate–adsorbent molecules is utilized which is based on their various thermodynamic bulk properties. A systematic approach to select these force fields is to match the corresponding experimental data with simulated vapor–liquid equilibrium properties [48]. The force fields commonly used for MOFs for carbon capture are UFF [49], GAFF [50], AMBER [51–53], TraPPE [48], DREIDING [54], OPLS force field [55, 56], CHARMM [57–59], MM3 [60], MM4 [61] JOYCE [62, 63] and CGenFF [64]. Dzubak et al. generated force field parameters for CO_2 adsorption in Mg-MOF-74 using the nonempirical model potential (NEMO) methodology to calculate the total interaction energy obtained from QM calculations from electrostatics, dispersion, repulsion and polarization. The parameters in each term of the force field were then fitted separately to generate the corresponding energy contributions [47]. Tafipolsky and Schmid used a genetic algorithm approach to derive force field parameters that include framework flexibility [65]. For clusters, combined quantum mechanics and molecular mechanics (QM/MM) methods can be used as well [66–73] with QM methods for the chemically reactive portion of the cluster and a set of structural and interaction parameters for the MM part. The QM/MM accuracy depends on the choice of the QM methodology and MM parameters and also on the treatment of electrostatics at the QM–MM boundary [74].

4.5.2 Grand Canonical Monte Carlo simulation

Grand canonical Monte Carlo (GCMC) simulations are widely used and the most convenient tool to study the adsorption properties of confined and bulk fluids [75, 76]. Monte Carlo simulations provide a powerful means for obtaining the temperature dependence of the binding energy of adsorbed molecules to their surface structures and their relative stability with density fluctuations which can be easily incorporated in the simulation procedure [14]. In such type of simulation, the volume, chemical potential and temperature are fixed while the number of molecules to be adsorbed is allowed to fluctuate [77, 78]. When the adsorbed phase is surrounded by gas, the temperature and chemical potential of the gas molecules inside and outside the adsorbent material are in equilibrium and each GCMC simulation consist of millions of random moves that sample the chosen ensemble [79]. Accurate description of the interatomic potential between the carbon dioxide molecules and the MOFs has to be determined [80]. For carbon dioxide, atomic point charge model (in electron units) can be assigned to the carbon C (+0.72) and the oxygen O (−0.36) atoms [46]. Metropolis scheme is commonly used with creation, rotation, deletion and displacement of fluid particles [81]. Ab initio methods or density functional methods can be used for GCMC simulation. Quantum mechanical calculations and charge equilibration methods can also be used to rapidly screen a larger number of MOFs for carbon dioxide capture at low pressures. These simulations agree well with experimental results and have identified a number of high-performance MOFs [43]. Generic force fields along with GCMC simulation were successfully used to study gas uptake in MOF-5 [40, 82, 83]. It is essential to account for the electrostatic interactions and van der Waals attraction and Morse function is the most suitable for studying gas adsorption in porous frameworks [84]. The numbers of the unit cells contained in the simulation box are MOF dependent, ranging from $1 \times 1 \times 1$ to $7 \times 7 \times 7$ so that enough molecules are accommodated to guarantee the simulation accuracy [85]. Kurniawan studied the solid–fluid interaction parameters by adopting the standard Lorentz–Berthelot for binary mixture of methane and carbon dioxide in slit-shaped carbon pores, ranging around 0.75–7.5 nm in width, for high pressure up to 300 bar and temperature range of 308–348 K [86]. Girardet et al. included point quadrupoles placed on the CO_2 molecules and made a model based on pairwise Lennard-Jones [87]. Christopher Daub et al. reported a monolayer adsorption for CO_2 adsorption on MgO sample by performing canonical ensemble of number of particles, volume, temperature and the grand canonical ensemble of viscosity, volume and temperature [14]. Deroche et al. computed the adsorption isotherms for AlPO-18 and STA-7 MOFs using a GCMC algorithm, implemented in the Accelerys Cerius2 programme [80]. Tenney and Lastoskie assigned partial charges and used Hartree–Fock method, polarization basis set 6–31 g(d,p) and Mulliken charge analysis with the Gaussian 03 software package [88]. Liu et al. and Lu et al. investigated the adsorption of CO_2 and mixed gases on heterogeneous surface models of coal, systematically through the density functional theory and GCMC [19, 89]. Vishnyakov et al. studied the adsorption of CO_2 in slit-shaped carbon micropores using GCMC and non-local density functional theory [90]. Pantatosaki et al. used GCMC to characterize microporous carbons and pore size distributions [91].

4.5.3 Computational Software

Cambridge Structural Database is a comprehensive database which has been used for studying and synthesizing MOFs but other materials which are not MOFs are also present in the database [92]. A database exclusively for MOFs is the online database COSMOS-ERC (https://cosmoserc.ku.edu.tr/) and the as-produced MOFs have excellent CO_2 separation capabilities. This database is useful for studying MOFs for CO_2/N_2, CO_2/CH_4 and $CO_2/N_2/CH_4$ separations and their structural properties. It is user friendly and it can be used to quantify structure–performance information to computationally propose novel MOFs [93]. CoRE MOF database (computation ready experimental MOFs) is also a widely used database for CO_2 separation from N_2 and CH_4 [94]. Quantitative relations between adsorption selectivities of MOFs and their metal types can be examined using this database [95].

4.6 Validation of simulation studies

Carbon capture experiments have to be performed on specified MOFs and the adsorption isotherms can be plotted with the help of temperature, heat of adsorption and pressure data. The simulated adsorption isotherms should be compared with experimental adsorption isotherms to validate the study. Adsorption and desorption isotherms can be obtained by simulation and these will be compared with experimental studies and it provides a means of judging the accuracy. Desorption is the opposite phenomenon of adsorption where the substance adsorbed is released from the adsorbent by varying the pressure or temperature. Care should be exercised while comparison between simulation and experiments in that the fitting should be done on a comprehensive experimental dataset rather than on isolated samples [96]. The differential enthalpy of adsorption over a range of pressure obtained from simulation and experiment can be compared to validate the inter atomic potential by calculating the isotherms and the evolution of the differential enthalpies of adsorption as a function of the pressure [97]. The excess adsorption per unit mass of carbon is a quantity useful for comparing theoretical with experimental values [98]. Most of the validation studies have been successful. Yazaydin et al. reported that relatively satisfactory predictions compared with experiments for most of the MOFs investigated under different pressures were obtained for 14 MOFs, including IRMOFs, HKUST-1 (or Cu-BTC),

ZIF-8, MOF-177, MIL-47 and the M-MOF-74 series [99]. Alsmail et al. investigated the use of computational chemistry and found that the simulation result was in excellent agreement with experimental data from the low-pressure region up to 20 bar for CH_4 and CO_2 adsorption in a new oxamide-containing MOF, NOTT-125a, at 298 K [100]. Wilmer et al. screened approximately 1,38,000 hypothetical MOFs for methane storage applications and experimentally confirmed their simulated predictions for a top-performing MOF [42]. Some cases where the results have been found to differ can be attributed to variations in geometrical and experimental surface areas and it is possible to adjust the simulation results by scaling the adsorbed quantities by an empirical factor [101, 102].

4.7 Summary

Computational chemistry is an indispensable tool in studying carbon capture by MOFs. Force fields have to be selected appropriately where principles of QM/MM can be used to devise the electrostatic charges and isotherm parameters on the MOFs. Gas adsorption isotherms have to be studied by GCMC simulation and compared with experimental adsorption isotherms for effective adsorption of CO_2. The available software packages in the market make the tedious process of screening MOFs less time consuming and give accurate information about the selectivity and working capacity which are necessary inputs for pressure swing adsorption. Computational chemistry is invaluable in calculating macroscopic and microscopic details of systems application in industries which are involved in carbon capture.

References

[1] IPCC. IPCC special report on carbon dioxide capture and storage. Cambridge: Cambridge University Press, 2005
[2] Furukawa H, Ko N, Go BY, Aratani N, Choi SB, Choi E, et al. Ultrahigh porosity in metal-organic frameworks. Science. 2010;329:424–8.
[3] Rowsell JL, Yaghi OM. Metal-organic frameworks: a new class of porous materials. Microporous Mesoporous Mater. 2004;73:3–14.
[4] Altintas C, Erucar I, Keskin S. High-throughput computational screening of the metal organic framework database for CH_4/H_2 separations. ACS Appl Mater Interfaces. 2018;10:3668–3679.
[5] Duren T, Bae YS, Snurr RQ. Using molecular simulation to characterize metal-organic frameworks for adsorption applications. Chem Soc Rev. 2009;38:1237–47.
[6] Zhang H, Bucior BJ, Snurr RQ. Molecular modeling of carbon dioxide adsorption in metal-organic frameworks. In Modelling and simulation in the science of micro-and meso-porous materials. USA: Elsevier, 2018:99–149. 9780128050576. 07.05.2019.
[7] Jose DF, Timothy F, Sean P, Howard M, Rameshwar DS. Advances in CO_2 capture technology—the U.S. department of energy's carbon sequestration program. Int J Greenhouse Gas Control. 2008;2:9–20.

[8] Orr JF. CO_2 capture and storage: are we ready? Energy Environ Sci. 2009;2:449–58.

[9] Jon G, Hannah C. Carbon capture and storage. Energy Policy. 2008;36:4317–22.

[10] Bourrelly S, Llewellyn PL, Serre C, Millange F, Loiseau T, Fe´rey GJ. Different adsorption behaviors of methane and carbon dioxide in the isotypic nanoporous metal terephthalates MIL-53 and MIL-47. Am Chem Soc. 2005;127:13519–21.

[11] Alessandro DM, Smit B, Jeffrey R. Long carbon dioxide capture: prospects for new materials. Int J Greenhouse Gas Control. 2008;2:9–20.

[12] Choi S, Drese JH, Jones CW. Adsorbent materials for carbon dioxide capture from large anthropogenic point sources. ChemSusChem. 2009;2:785–891.

[13] Daub CD, Patey GN. Monte Carlo simulations of the adsorption of CO_2 on the MgO(100) surface. J Chem Phys. 2006;124:1–9.

[14] Lithoxoos GP, Labropoulos A, Peristeras LD, Kanellopoulos N, Samios J, Ioannis GE. Adsorption of N_2, CH_4, CO and CO_2 gases in single walled carbon nanotubes: a combined experimental and Monte Carlo molecular simulation study. J Supercrit Fluids. 2010;55:510–23.

[15] Bulut E, Ozacar M, Sengil IA. Adsorption of malachite green onto bentonite: equilibrium and kinetic studies and process design. Micropor Mesopor Mater. 2008;115:234–46.

[16] Malek A, Farooq S. Comparison of isotherm models for hydrocarbon adsorption on activated carbon. AIChE J. 1996;42:3191–201.

[17] Ayawei N, Ebelegi AN, Wankasi D. Modelling and interpretation of adsorption isotherms. J Chem. 2017;Article ID 3039817:1–11.

[18] Sudik AC, Millward AR, Ockwig NW, Côté AP, Kim J, Yaghi OM. Design, synthesis, structure, and gas (N_2, Ar, CO_2, CH_4, and H_2) sorption properties of porous metal-organic tetrahedral and heterocuboidal polyhedra. J Am Chem Soc. 2005;127:7110–8.

[19] Liu XQ, He X, Qiu NX, Yang X, Tian ZY, Li MJ. Molecular simulation of CH_4, CO_2, H_2O and N_2 molecules adsorption on heterogeneous surface models of coal. Appl Surf Sci. 2016;389:894–905.

[20] Pan Y, Li Z, Zhang Z, Tong XS, Li H, Jia CZ, et al. Adsorptive removal of phenol from aqueous solution with zeolitic imidazolate framework-67. J Environ Manage. 2016;169:167–73.

[21] Meek ST, Greathouse JA, Allendorf MD. Metal-Organic frameworks: a rapidly growing class of versatile nanoporous materials. Adv Mater. 2011;23:249–67.

[22] Allendorf M, Bauer C, Bhakta R, Houk R. Luminescent metal–organic frameworks. Chem Soc Rev. 2009;38:1330–52.

[23] Murray LJ, Dincă M, Long JR. Hydrogen storage in metal–organic frameworks. Chem Soc Rev. 2009;38:1294–314.

[24] Millward AR, Yaghi OM. Metal-organic frameworks with exceptionally high capacity for storage of carbon dioxide at room temperature. J Am Chem Soc. 2005;127:17998–9.

[25] Czaja AU, Trukhan N, Muller U. Industrial applications of metal-organic frameworks. Chem Soc Rev. 2009;38:1284–93.

[26] Keskin S, van Heest TM, Sholl DS. Can metal–organic framework materials play a useful role in large-scale carbon dioxide separations? ChemSusChem. 2010;3:879–91.

[27] Lee JY, Farha OK, Roberts J, Scheidt KA, Nguyen SB, Hupp JT. Metal–organic framework materials as catalysts. Chem Soc Rev. 2009;38:1450–9.

[28] Horcajada P, Chalati T, Serre C, Gillet B, Sebrie C, Baati T, et al. Porous metal-organic-framework nanoscale carriers as a potential platform for drug delivery and imaging. Nat Mater. 2009;9:172–8.

[29] Ben-Mansour R, Habib M, Bamidele O, Basha M, Qasem N, Peedikakkal A, et al. Carbon capture by physical adsorption: materials, experimental investigations and numerical modelling and simulations–a review. Appl Energy. 2016;161:225–255.

[30] Eddaoudi M, Moler DB, Li H, Chen B, Reineke TM, Keeffe MO, et al. Modular chemistry: secondary building units as a basis for the design of highly porous and robust metal–organic carboxylate frameworks. Acc Chem Res. 2001;34:319–30.
[31] Tranchemontagne DJ, Mendoza-Cortes JL, Keeffe MO, Yaghi OM. Secondary building units, nets and bonding in the chemistry of metal–organic frameworks. Chem Soc Rev. 2009;38:1257–83.
[32] Yang QY, Zhong CL, Chen JF. Computational study of CO_2 storage in metal–organic frameworks. J Phys Chem C. 2008;112:1562–9.
[33] Schoedel A, Ji Z, Yaghi OM. The role of metal–organic frameworks in a carbon-neutral energy cycle. Nat Energy. 2016;1:16034.
[34] Caskey SR, Wong-Foy AG, Matzger AJ. Dramatic tuning of carbon dioxide uptake via metal substitution in a coordination polymer with cylindrical pores. J Am Chem Soc. 2008;130:10870–1.
[35] Rosi NL, Kim J, Eddaoudi M, Chen B, Keeffe MO, Yaghi OM. Rod packings and metal–organic frameworks constructed from rod-shaped secondary building units. J Am Chem Soc. 2005;127:1504–18.
[36] Nugent P, Belmabkhout Y, Burd SD, Cairns AJ, Luebke R, Forrest K, et al. Porous materials with optimal adsorption thermodynamics and kinetics for CO_2 separation. Nature. 2013;495:80–4.
[37] Li B, Zhang Z, Li Y, Yao K, Zhu Y, Deng Z, et al. Enhanced binding affinity, remarkable selectivity, and high capacity of CO_2 by dual functionalization of a *rht*-type metal–organic framework. Angew Chem Int Ed. 2012;51:1412–15.
[38] Cheon YE, Park J, Suh MP. Selective gas adsorption in a magnesium-based metal–organic framework. Chem Commun. 2009;36:5436–8.
[39] Keith DW, Minh HD, Stolaroff JK. Climate strategy with Co_2 capture from the air. Clim Change. 2006;74:17–45.
[40] Yang Q, Zhong C. Molecular simulation of carbon dioxide/methane/hydrogen mixture adsorption in metal–organic frameworks. J Phys Chem B. 2006;110:17776–83.
[41] Arstad B, Fjellvåg H, Kongshaug KO, Swang O, Blom R. Amine functionalised metal organic frameworks (MOFs) as adsorbents for carbon dioxide. Adsorption. 2008;14:755–762.44.
[42] Wilmer CE, Leaf M, Lee CY, Farha OK, Hauser BG, Hupp JT, et al. Large-scale screening of hypothetical metal–organic frameworks. Nat Chem. 2012;4:83–9.
[43] Loisruangsin A, Fritzsche S, Hannongbua S. Newly developed ab initio fitted potentials for molecular dynamics simulations of *n*-pentane in the zeolite silicalite-1.S. Chem Phys Lett. 2004;390:485–90.
[44] Maurin G, Llewellyn PL, Poyet T, Kuchta B. Influence of extra-framework cations on the adsorption properties of X-Faujasite systems: microcalorimetry and molecular simulations. J Phys Chem B. 2005;109:125–9.
[45] Schmidt JR, Yu K, McDaniel JG. Transferable next-generation force fields from simple liquids to complex materials. Acc Chem Res. 2015;48:548–56.
[46] Dzubak AL, Lin LC, Kim J, Swisher A, Poloni R, Maximoff SN, et al. Ab initio carbon capture in open-site metal-organic frameworks. Nat Chem. 2012;4:810–16.
[47] Vanduyfhuys L, Vandenbrande S, Verstraelen T, Schmid R, Waroquier M, Speybroeck VV. QuickFF: a program for a quick and easy derivation of force fields for metal-organic frameworks from Ab initio input. J Comput Chem. 2015;13:1015–27.
[48] Potoff JJ, Siepmann JI. Vapor–liquid equilibria of mixtures containing alkanes, carbon dioxide, and nitrogen. J I AIChE J. 2001;47:1676–82.
[49] Rappe AK, Casewit CJ, Colwell KS, Goddard III WA, Skiff WM. UFF, a full periodic table force field for molecular mechanics and molecular dynamics simulations. J Am Chem Soc. 1992;114:10024–35.

[50] Wang J, Wolf R, Caldwell J, Kollman P, Case D. Development and testing of a general amber force field. J Comput Chem. 2004;25:1157–74.

[51] Weiner S, Kollman P, Nguyen D, Case D, Singh U, Ghio C, et al. A new force field for molecular mechanical simulation of nucleic acids and proteins. J Am Chem Soc. 1984;106:765–84.

[52] Weiner S, Kollman P, Nguyen D, Case D. An all atom force field for simulations of proteins and nucleic acids. J Comput Chem. 1986;7:230–52.

[53] Cornell W, Cieplak P, Bayly C, Gould I, Merz K, Ferguson D, et al. A second generation force field for the simulation of proteins, nucleic acids, and organic molecules. J Am Chem Soc. 1995;117:5179–97.

[54] Mayo SL, Olafson BD, Goddard WA. DREIDING—a generic force-field for molecular simulations. J Phys Chem. 1990;94:8897–909.

[55] Jorgensen WL, Tirado-Rives J. The OPLS [optimized potentials for liquid simulations] potential functions for proteins, energy minimizations for crystals of cyclic peptides and crambin. J Am Chem Soc. 1988;110:1657–66.

[56] Jorgensen WL, Maxwell DS, Tirado-Rives J. Development and testing of the OPLS all-atom force field on conformational energetics and properties of organic liquids. J Am Chem Soc. 1996;118:11225–36.

[57] MacKerell AD, Bashford D, Bellott M, Dunbrac RL, Evanseck JD, Field MJ, et al. All-atom empirical potential for molecular modelling and dynamics studies of proteins. J Phys Chem B. 1998;102:3586–616.

[58] MacKerell AD, Banavali N, Foloppe N. Development and current status of the CHARMM force field for nucleic acids. Biopolymers. 2001;56:257–65.

[59] MacKerell AD, Feig M, Brooks C. Extending the treatment of backbone energetics in protein force fields: Limitations of gas-phase quantum mechanics in reproducing protein conformational distributions in molecular dynamics simulations. J Comput Chem. 2004;25:1400–15.

[60] Allinger NL, Yuh YH, Lii JH. Molecular mechanics. The MM3 force field for hydrocarbons. 1. J Am Chem Soc. 1989;111:8551–66.

[61] Allinger NL, Chen K, Lii JH. An improved force field (MM4) for saturated hydrocarbons. J Comput Chem. 1996;17:642–68.

[62] Cacelli I, Prampolini G. Parametrization and validation of intramolecular force fields derived from DFT calculations. J Chem Theory Comput. 2007;3:1803–17.

[63] Barone V, Cacelli I, De Mitri N, Licari D, Monti S, Prampolini G. JOYCE and ULYSSES: integrated and user-friendly tools for the parameterization of intramolecular force fields from quantum mechanical data. Phys Chem Chem Phys. 2013;15:3736–51.

[64] Vanommeslaeghe K, Hatcher E, Acharya C, Kundu S, Zhong S, Shim J, et al. CHARMM general force field: a force field for drug-like molecules compatible with the CHARMM all-atom additive biological force fields. J Comput Chem. 2010;31:671–90.

[65] Tafipolsky M, Schmid R. Systematic first principles parameterization of force fields for metal–organic frameworks using a genetic algorithm approach. J Phys Chem B. 2009;113:1341–52.

[66] Hirao H, Ng WK, Moeljadi AM, Bureekaew S. Multiscale model for a metal–organic framework: high-spin rebound mechanism in the reaction of the oxoiron(IV) species of Fe-MOF-74. ACS Catal. 2015;5:3287–3291.

[67] Choomwattana S, Maihom T, Khongpracha P, Probst M, Limtrakul J. Structures and mechanisms of the carbonyl-ene reaction between MOF-11 encapsulated formaldehyde and propylene: an ONIOM study. J Phys Chem C. 2008;112:10855–10861.

[68] Doitomi K, Xu K, Hirao H. The mechanism of an asymmetric ring-opening reaction of epoxide with amine catalyzed by a metal–organic framework: insights from combined quantum mechanics and molecular mechanics calculations. Dalton Trans. 2017;46:3470–3481.

[69] Oxford GA, Snurr RQ, Broadbelt LJ. Hybrid quantum mechanics/molecular mechanics investigation of (salen) Mn for use in metal–organic frameworks. Ind Eng Chem Res. 2010;49:10965–10973.

[70] Zheng M, Liu Y, Wang C, Liu S, Lin W. Cavity-Induced enantioselectivity reversal in a chiral metal–organic framework brønsted acid catalyst. Chem Sci. 2012;3:2623–2627.

[71] Yu D, Yazaydin AO, Lane JR, Dietzel PD, Snurr RQ. A combined experimental and Quantum chemical study of CO_2 adsorption in the metal–organic framework CPO-27 with different metals. Chem Sci. 2013;4:3544–3556.

[72] Moeljadi AM, Schmid R, Hirao H. Dioxygen binding to Fe-MOF-74: microscopic insights from periodic QM/MM calculations. Can J Chem. 2016;94:1144–1150.

[73] Doitomi K, Hirao H. Hybrid computational approaches for deriving quantum mechanical insights into metal–organic frameworks. Tetrahedron Lett. 2017;58:2309–2317.

[74] Pezeshki S, Lin H. Recent developments in QM/MM methods towards open-boundary multi-scale simulations. Mol Simul. 2015;41:168–89.

[75] Allen FH. The Cambridge structural database: a quarter of a million crystal structures and rising. Acta Crystallogr B. 2002;58:380–8.

[76] Bhatia SK, Myers AL. Optimum conditions for adsorptive storage. Langmuir. 2006;22:1688–700.

[77] Frenkel D, Smit B. Understanding molecular simulation: from algorithms to applications. 2nd ed. San Diego: Academic Press, 2002.

[78] Adatoz E, Keskin S. Application of MD simulations to predict membrane properties of MOFs. J Nanomater. 2015;1:1–9.

[79] Duren T, Bae YS, Randall Q. Snurr using molecular simulation to characterize metal–organic frameworks for adsorption applications. Chem Soc Rev. 2009;38:1237–47.

[80] Deroche I, Gaberova L, Maurin G, Llewellyn P, Castro M, Wright P. Adsorption of carbon dioxide in SAPO STA-7 and AlPO-18: grand Canonical Monte Carlo simulations and microcalorimetry measurements. Adsorption. 2008;14:207–13.

[81] Metropolis N, Rosenbluth AW, Rosenbluth MN, Teller AN, Teller E. Equation of State Calculations by Fast Computing Machines. J Chem Phys. 1953;21:1087–92.

[82] Duren T, Sarkisov L, Yaghi OM, Snurr RQ. Design of new materials for methane storage. Langmuir. 2004;20:2683–9.

[83] Jhon YH, Cho M, Jeon HR, Park I, Chang R, Rowsell JLC, et al. Simulations of methane adsorption and diffusion within alkoxy-functionalized IRMOFs exhibiting severely disordered crystal structures. J J Phys Chem C. 2007;111:16618–25.

[84] Jose L, Corte M, Han SS, Furukawa H, Yaghi OM, Goddard WA. Adsorption mechanism and uptake of methane in covalent organic frameworks: theory and experiment. J Phys Chem A. 2010;114:10824–33.

[85] Wu D, Wang C, Liu B, Liu D, Yang Q, Zhong C. Large-scale computational screening of metal-organic frameworks for CH_4/H_2 separation. AIChE J. 2011;58:2078–84.

[86] Kurniawan Y, Bhatia SK, Rudolph V. Monte Carlo simulation of binary mixture adsorption of methane and carbon dioxide in carbon slit pores. Technical report, University of Queensland, 2005.

[87] Girardet C, Picaud S, Hoang PN. A comparitive study of the geometry of CO_2 monolayers adsorbed on the ionic substrates NaCl and MgO(100). Europhys Lett. 1994;25:131–6.

[88] Tenney CM, Lastoskie CM. Molecular simulation of carbon dioxide adsorption in chemically and structurally heterogeneous porous carbons. Environ Progress. 2006;25:343–54.

[89] Lu X, Jin D, Wei S, Zhang M, Zhu Q, Shi X. Competitive adsorption of a binary CO_2-CH_4 mixture in nanoporous carbons: effects of edge-functionalization. Nanoscale. 2015;7:1002–12.

[90] Vishnyakov A, Ravikovitch PI, Neimark AV. Molecular level models for CO_2 sorption in nanopores. Langmuir. 1999;15:8736–42.

[91] Pantatosaki E, Psomadopoulos D, Steriotis T, Stubos AK, Papaioannou A, Papadopoulos GK. Micropore size distributions from CO_2 using grand canonical Monte Carlo at ambient temperatures: cylindrical versus slit pore geometries. Colloids Surf A: Physicochem Engng Aspects. 2004;241:127–35.

[92] Moghadam PZ, Li A, Wiggin SB, Tao A, Maloney AG, Wood PA, et al. Development of a cambridge structural database subset: a collection of metal-organic frameworks for past, present, and future. Chem Mater. 2017;29:2618–2625.

[93] Altintas C, Avci G, Daglar H, Azar AN, Velioglu S, Erucar I, et al. Database for CO_2 separation performances of MOFs based on computational materials screening. ACS Appl Mater Interfaces. 2018;10:17257–17268.

[94] Chung YG, Camp J, Haranczyk M, Sikora BJ, Bury W, Krungleviciute V, et al. Computation-ready, experimental (CoRE) metal-organic frameworks: a tool to enable high-throughput screening of nanoporous crystals. Chem Mater. 2014;26:6185–92.

[95] Qiao Z, Zhang K, Jiang J. In silico screening of 4764 computation-ready, experimental metal–organic frameworks for CO_2 separation. J Mater Chem A. 2016;4:2105–2114.

[96] Coudert FX, Fuchs AH. Computational characterization and prediction of metal–organic frame work properties. Coord Chem Rev. 2016;307:211–36.

[97] Gu SW, Gao B, Teng L, Li Y, Fan C, Iglauer S, et al. Monte Carlo simulation of supercritical carbon dioxide adsorption in carbon slit pores. Energy Fuels. 2017;31:9717–24.

[98] Zhao A, Samanta A, Sarkar P, Gupta R. Carbon dioxide adsorption on amine-impregnated mesoporous SBA-15 sorbents: experimental and kinetics study. Ind Eng Chem Res. 2013;52:6480–91.

[99] Yazaydın AO, Snurr RQ, Park TH, Koh K, Liu J, LeVan MD, et al. Screening of metal–organic frameworks for carbon dioxide capture from flue gas using a combined experimental and modelling approach. J Am Chem Soc. 2009;131:18198–9.

[100] Alsmail NH, Suyetin M, Yan Y, Cabot R, Krap CP, Lü J, et al. Analysis of high and selective uptake of CO_2 in an oxamide containing $\{Cu_2(OOCR)_4\}$-based metal-organic framework. Chem Eur J. 2014;20:7317–24.

[101] Sillar K, Sauer J. Ab initio prediction of adsorption isotherms for small molecules in metal–organic frameworks: the effect of lateral interactions for methane/CPO-27-Mg. J Am Chem Soc. 2012;134:18354–65.

[102] Fairen-Jimenez D, Lozano-Casal P, Duren T, Kaskel S, Llewellyn P, Rodriguez-Reinoso F, et al., editors. Proceedings of the 8th international symposium on the characterisation of porous solids. Royal Soc Chem. 2009;119:80–7.

Varun Chahal, Sonam Nirwan and Rita Kakkar

5 Combined approach of homology modeling, molecular dynamics, and docking: computer-aided drug discovery

Abstract: With the continuous development in software, algorithms, and increase in computer speed, the field of computer-aided drug design has been witnessing reduction in the time and cost of the drug designing process. Structure based drug design (SBDD), which is based on the 3D structure of the enzyme, is helping in proposing novel inhibitors. Although a number of crystal structures are available in various repositories, there are various proteins whose experimental crystallization is difficult. In such cases, homology modeling, along with the combined application of MD and docking, helps in establishing a reliable 3D structure that can be used for SBDD. In this review, we have reported recent works, which have employed these three techniques for generating structures and further proposing novel inhibitors, for *cytoplasmic proteins*, *membrane proteins*, and *metal containing proteins*. Also, we have discussed these techniques in brief in terms of the theory involved and the various software employed. Hence, this review can give a brief idea about using these tools specifically for a particular problem.

Keywords: homology modeling, docking, molecular dynamics, cytoplasmic proteins, membrane proteins, metalloproteins

5.1 Introduction

Computer-aided drug design (CADD) is a fast growing field, and with advances in the various approaches used, the time and cost in developing novel drugs have been greatly reduced. These approaches include ligand based virtual screening, ligand based pharmacophore modeling [1–3], structure based virtual screening and pharmacophore modeling [4], and many others. Among these approaches, structure based drug designing (SBDD) requires the 3D structure of the protein, but, currently, the 3D structure information can be generated for only up to 56% of all known proteins. Thus, homology modeling or comparative modeling has been developed in the recent past for developing the 3D structure of proteins computationally [5]. Although this technique turns out to be helpful for the proteins whose

This article has previously been published in the journal Physical Sciences Reviews. Please cite as: Chahal, V., Nirwan, S., Kakkar, R. Combined approach of homology modeling, molecular dynamics, and docking: computer-aided drug discovery *Physical Sciences Reviews* [Online] 2020, 5 DOI: 10.1515/psr-2019-0066.

https://doi.org/10.1515/9783110631623-005

experimental structure is either not available or difficult to deposit, the quality of the predicted model needs to be refined and validated using molecular dynamics (MD) and molecular docking approaches, respectively. Thus, in this review, we have reported studies where the combined approach of homology modeling, MD and molecular docking has been employed (Figure 5.1).

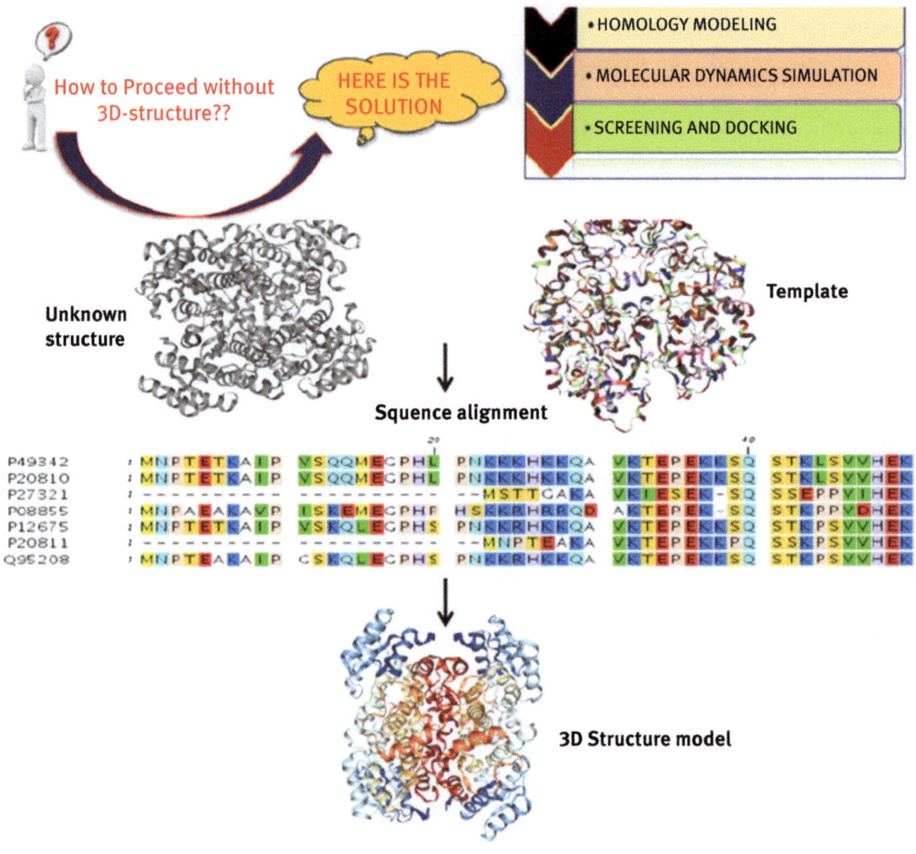

Figure 5.1: Combined approach of homology modeling, molecular dynamics, and docking in SBDD.

We have reported recent work that employs these three techniques in conjugation for proposing 3D structures of different types of proteins and their novel inhibitors. Proteins such as membrane proteins, metalloproteinase, and cytoplasmic proteins are reviewed. Brief descriptions of each of these techniques have also been given, providing a basic understanding of the techniques, along with the software and tools required. Thus, this work can help in understanding these methodologies, along with their merits and shortcomings.

5.2 Homology modeling

In the absence of a three dimensional protein structure, homology modeling is a powerful tool to predict the 3D structures of proteins [6–8]. This methodology is based on the general principle that proteins with similar sequences have similar tertiary structures. Recently, a number of programs and servers such as MODELLER, SWISS-MODEL, RAMP, PrISM, COMPOSER, CONGEN + 2, and DISGEO/Consensus have been developed to build a complete 3D model. The process of homology modeling mainly consists of four steps, as shown in Figure 5.2 [9]:

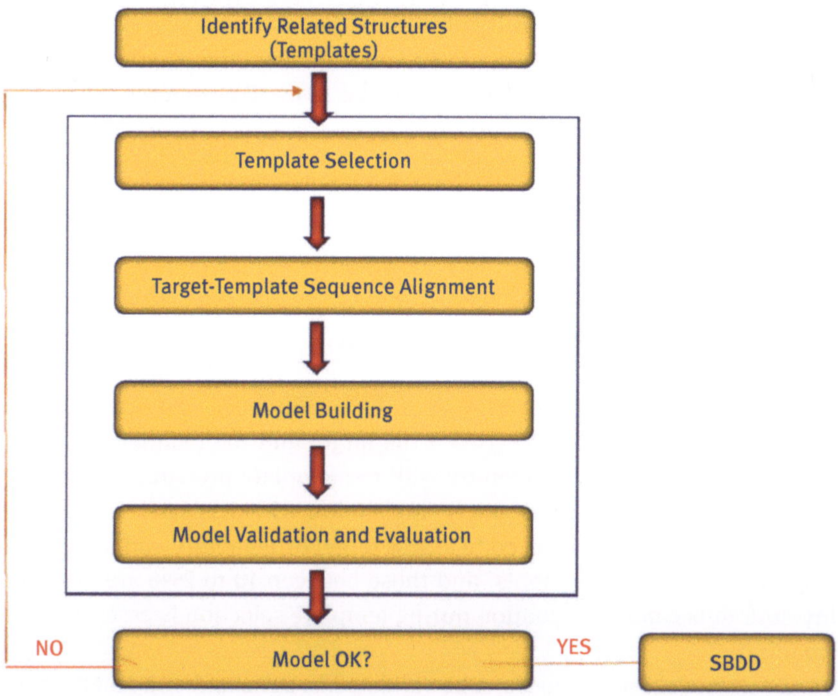

Figure 5.2: Steps involved in homology modeling for 3D structure prediction.

These steps are briefly as follows:

1. *Template selection or recognition*

In this initial step, the 2D sequence of the target protein is compared with the sequences of proteins with known 3D structures available in various depositories such as PDB, using the BLAST server. Various repositories such as SwissProt database and UniProtKB are used for availing the required primary protein sequence for the target protein. The suitable protein thus found, also known as the template protein, serves as a template in model building. In some cases, a single template is not enough to provide the complete structural information for the target protein and, therefore, the multiple template approach [10], or mix and match technique [11], has also been used for improving the overall model quality.

2. *Single or multiple sequence alignments of target and template proteins*
Sequence alignment plays a major role for the development of an accurate homology model [5]. For a given biological sequence, the main purpose of a multiple sequence alignment (MSA) method is to align the sequences in a way that will show their evolutionary, functional or structural relationship. This is done by allowing the homologous positions to be aligned with each other and inserting gaps of varying length within the sequences. For a structural model, the aligned residues should have comparable positions in their respective 2D or 3D structures [12].

3. *Model building*
A number of methods like rigid-body assembly, segment matching, spatial restraint and artificial evolution are used for model building. The accuracy of the model depends on the template selection and alignment process.

4. *Evaluation, validation, and refinement of the model*
These three steps, mentioned above, should be repeated multiple times until a satisfactory model is obtained. As a general rule, the quality of the model is directly linked with the sequence identities between the target and the template. A model based on 50% or greater sequence identity with the template protein is considered to be accurate enough for drug discovery applications. Models having sequence identity between 25 to 50% can be used to assess target druggability and are helpful in designing mutagenesis experiments, and those between 10 to 25% are of uncertain quality, and thus careful evaluation during template selection is essential for a reliable model.

The proposed model further needs to be validated in terms of bond length, bond angles, and steric clashes between the side chains of residues. There are various tools employed for this purpose, such as PROCHECK [13], ProSA [14], VERIFY-3D, PDBeFOLD [15], and CHIMERA [16]. PROCHECK generates Ramachandran plots and G-factor values which inspect the stereochemical quality of the models by assessing whether the amino acids are in a phi-psi distribution that is consistent with a right-handed α-helix. In order to investigate whether the interaction energy of each residue with the remainder of the protein is negative, a second test using the ProSA energy plot is performed. Figure 5.3 shows the ProSA plot [17] (3a) and Ramachandran plot (3b) [18] for the two different homology modeled protein structures.

Other techniques such as VERIFY-3D compare the model with its own template structure in terms of three-dimensional and sequence score (3D–1D score). A score

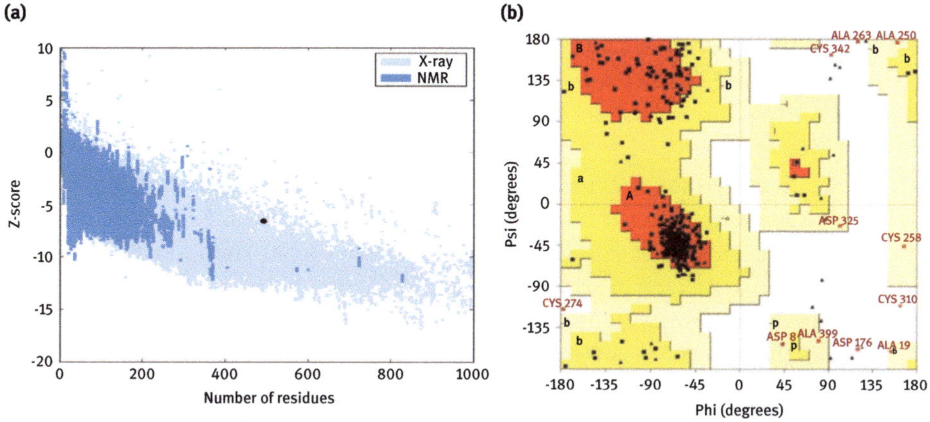

Figure 5.3: ProSA plot (3a) and Ramachandran plot (3b) of homology modeled protein showing overall quality of the modeled structure. (Reprinted with permission).

above zero denotes good compatibility between the 3D and 2D forms. Furthermore, superposition of the modeled protein with the 3D template structure gives the root mean square deviation (RMSD) factor, and a lower RMSD suggests good superposition. Various online tools like PDBeFOLD, along with CHIMERA, give the required C-alpha backbone RMSD. The model is considered suitable if it passes these verification parameters; otherwise, further refinement is needed. This can be done either by using the loop refinement option in MODELER or by performing MD simulation, discussed in brief in the next section.

5.3 Molecular dynamics

MD is a computational simulation method, used to study the interaction and motion of atoms and molecules according to Newton's physics. Atom and molecules are allowed to interact for a fixed period of time. The integration of Newton's laws of motions for a system of interacting particles gives successive configurations of the system that provide trajectories, specifying positions and velocities of the particles over time. The forces between interacting particles and the overall energy of the system are calculated using force fields. Force fields are mathematical equations that are easy to calculate and describe the dependence of the energy of a system on the coordinates of its particles. The most popular force fields used are AMBER, CHARMM, GROMOS, OPLS and COMPASS. Out of these five, the first three are generally used in simulation of biomolecules. Different forms of these force fields are continuously developing; for example, CHARMM19, CHARMM22, CHARMM27, GROMOS96, GROMOS45A3, GROMOS53A5, GROMOS53A6, AMBER91, AMBER94, AMBER96,AMBER99, AMBER02, etc. [19].

Nowadays, MD is emerging as a routine tool in the field of CADD, which employs either the three-dimensional structures of the protein (SBDD) or known bioactive ligands (ligand based drug designing) to discover promising candidate drugs. At present, the protein data bank (PDB) holds more than 100,000 proteins for the SBDD workflow. However, structures stored in the PDB represent only a partial view of the 3D structures, whereas proteins are flexible entities and undergo significant conformational changes while performing their functions. Especially their active sites are regions of both low and high conformational stability. Thus, in this regard, MD emerges as the most versatile technique that helps in understanding the dynamic behavior of proteins at different timescales [20].

Furthermore, in the combined approach with homology modeling, where 3D structures are proposed, MD simulation on the proposed structure minimizes various steric clashes between the side chains, and hence a model close to the experimental structure is obtained. Also, in cases where novel inhibitors are proposed through CADD, MD simulations on the protein-ligand complex can show the binding of the ligand over the time period.

5.3.1 MD on membrane proteins

Membrane proteins are responsible for the interactions of cells with their surrounding environment and constitute approximately 50% of the current drug targets. The generation of the lipid bilayer is the crucial step for their simulation. Several methods have been proposed for the construction of the protein-membrane system [21, 22]. A starting structure can be built by merging the protein structure with an existing equilibrated membrane patch. However, this method works well only for a membrane consisting of a single lipid type and also if the equilibrated bilayers are available. On the other hand, a bilayer consisting of multiple components requires re-equilibration, which may take up to microseconds of simulations and this is computationally very expensive. To overcome these problems, a variety of coarse gain (CG) models have been introduced that enable mesoscopic simulations on a multi-microsecond scale [23]. The most popular tool *insane* uses coarse-gain lipid templates for membrane building and provides an efficient means for generating equilibrated atomistic models for multi-component membranes [24].

5.4 Molecular docking

After a valid model has been prepared using homology modeling and further minimized using MD, the technique of molecular docking is employed for validating the binding modes in the active site of the modeled protein. Firstly, the active site of the modeled protein is located using the CASTp tool [25] or the SiteMap module of

Schrödinger. Then a ligand library of known inhibitors of the same class of enzymes/proteins is used for docking into this active site. Various docking tools have been employed for the purpose, such as Swiss Dock, Glide docking in Schrödinger [26], Auto Dock [27], and Automatic docking [28].

5.5 Applications

5.5.1 Cytoplasmic proteins

5.5.1.1 Cytochrome P450

The cytochrome P450s (CYPs) are heme containing proteins that are present in both eukaryotes and prokaryotes. The CYP enzymes catalyze a number of reactions that are involved in oxidative metabolism and also metabolize a wide variety of endogenous compounds like steroids, fatty acids and prostaglandins. The CYPs also act on exogenous substrates, including drugs, carcinogens and environment pollutants [29]. At present, 57 human functional CYP genes are known. The human CYP genes are classified into 18 families and 44 subfamilies [30] based on the amino acid sequence of the encoded protein. The heme containing active site of CYPs is highly flexible and undergoes conformational changes upon ligand binding. As noted above, CYPs are the enzymes involved in drug metabolism and thus represent a major area of research interest. In recent years, a plethora of work has been done on CYPs to understand their 3D structure and also to study the conformational changes in the active site.

Recently, a computational study on CYP24A1 was performed to identify CYP24A1 inhibitors, using various computational techniques like homology modeling, docking, MD simulations, 3D QSAR and pharmacophore modeling. CYP24A1 is a CYP 24-hydroxylase enzyme that plays a crucial role in maintaining the circulatory level of Vitamin D. Over-expression of CYP24A1 leads to deregulation of Vitamin D and results in several clinical pathogeneses like chronic kidney disease, osteoporosis, various forms of cancers, inflammatory functions, and type I diabetes [31–33]. In the investigation [34], due to the non-availability of the crystal structure for the human CYP24A1 protein, a homology model of human CYP24A1 was built and the crystal structure of the rat enzyme isoform of CYP24A1 was used as the template. The template and the target sequences had an 85% sequence identity. The quality of the model was verified by PROCHECK and ERRAT. Molecular docking studies on the CYP24A1 model resulted in three lead compounds. Furthermore, MD simulation studies showed that the compounds are stable and form many hydrogen bond interactions with the CYP24A1 protein and thus have potent inhibitory effect on the CYP24A1 protein.

Fan et al. [35] investigated the ligand selectivity in CYP3A7 protein using the combined approach of homology, docking, MD and MM-GB/SA. In their study, a 3D model of CYP3A7 was proposed using the crystal structure of CYP3A4 as the template. The constructed model was refined by MD simulations and then three ligands, viz. dehydroepiandrosterone (DHEA), estrone and estradiol were docked into the active site of the proposed 3D model. Results from docking, MD, along with MM-GB/SA studies, revealed that several residues like Phe108, Ser119, Phe304, Ala370 and Leu482 play important roles in ligand binding. Among these ligands, DHEA, which shows higher binding affinity, is stabilized by the formation of two hydrogen bonds with Ser119 and Arg372.

O. felineus CYP450 (OfCYP450), another important enzyme of the CYPs family, is considered as a promising drug target for the development of therapeutic agents against trematodiasis [36, 37]. In a recent work [38], a computational study involving different methodologies for identifying natural inhibitors of OfCYP450, has been reported. It involves structure based virtual screening of natural compounds from the ZINC database against homology model of OfCYP450, followed by MD simulations. The 3D structure was predicted using I-TASSER, an *ab initio* server. Furthermore, MD simulation over a time period of 50 ns was run to obtain the lowest energy structure. Results from the screening process yielded four potential inhibitors: ZINC2358298, ZINC8790946, ZINC70707116, and ZINC85878789. The inhibitors were further subjected to MD studies and the ligands ZINC8790946, ZINC70707116, and ZINC85878789 were found to form stable complexes with OfCYP450.

CYP7B1 is a steroid CYP 7α-hydroxylase that is widely expressed in the brain and also in the liver and kidney, albeit at a much lower level. A series of investigations including homology modeling, automatic docking, and MD have been reported for this protein. The major goal was to investigate the structural features needed for the substrate selectivity of CYP7B1. After proposing the model and performing docking using four substrates (25-HOChol, DHEA, anediol, and enediol), MD simulations were performed. It was found that the Phe cluster residues that lie above the active site, particularly Phe489, merge the active site with the adjacent channel to the surface and accommodate the substrate (25-HOChol) selectively [29].

5.5.1.2 Kinases

Kinases belong to a large family of phosphotransferases, and are the enzymes that catalyze the transfer of phosphate groups from high-energy, phosphate-donating molecules to specific substrates. They are further classified into various types on the basis of the substrates they act upon, such as *protein kinases* [39], *lipid kinases* [40] and *carbohydrate kinases* [41]. In this section, we report recent computational studies on various kinases that aim at studying their structure, along with developing novel inhibitors.

FMS-like tyrosine kinase 3 (FLT3), a type III receptor tyrosine kinase, has emerged as an alternative to traditional chemotherapy for the treatment of acute myeloid leukemia (AML). Nowadays, inhibition of FLT3 activity by small drug like molecules is considered as a novel treatment for AML patients [42, 43]. A number of FLT3 inhibitors are known that bind to the ATP binding site. However, there are some differences in their binding modes, depending upon the orientation of the DGF (Asp-Phe-Glu) motif in the activation loop. When the phenyl group of residue Phe (Phe830) is oriented outside the ATP binding site, the DGF motif adopts an "in" conformation, while in case of inside orientation of the phenyl group, it adopts the "out" conformation (Figure 5.4) [44]. It has been reported earlier that the inhibitors that bind to the "in conformation" mode are more promising for the treatment of AML.

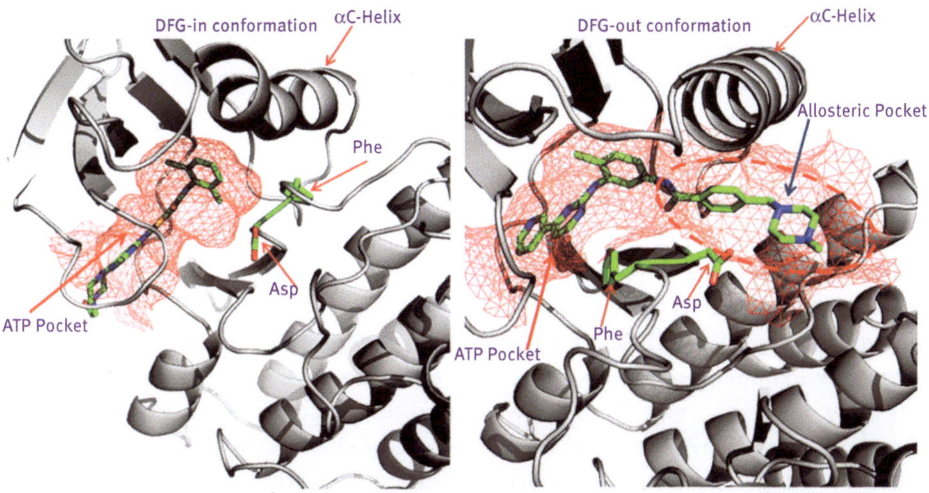

Figure 5.4: Two conformations adopted by kinases: the orientation of the residue Phe830 results in DFG-in conformation (if oriented away from ATP binding site) and DFG-out (if oriented inside the ATP binding site). (Reprinted with permission).

In this direction, Ke et al. [45] reported structure based screening of DFG-in FLT3 to identify its inhibitors. The 3D structure of DGF-in FLT3 was proposed by homology modeling, using DGF-out FLT3 (PDB: 1RJB) and DGF-in CSFR-1 (PDB: 3LCD) as templates (Figure 5.5). DGF-out FLT3 complex is the primary template whose DGF-out motif is replaced by the DGF-in motif of the second template, 3LCD. The 1RJB structure shares a sequence identity of 93% and the 3LCD structure shares a sequence identity of 62.7% with the target sequence. The overall quality factor of the modeled structure was found to be 84.53%. The built model was further refined by MD simulations. In the MD-simulated structure, 90% of the residues were found to be located in the most favored region and only 0.8% in the disallowed region. A ligand library of 125,000 compounds of an in-house HTS database were screened over the modeled structure and in this process two hits (BPR056 and BPR080) were identified. Furthermore, results from docking, DFT, and 20-ns MD simulation studies of these two hits showed that BPR056 forms a stable interaction with the target.

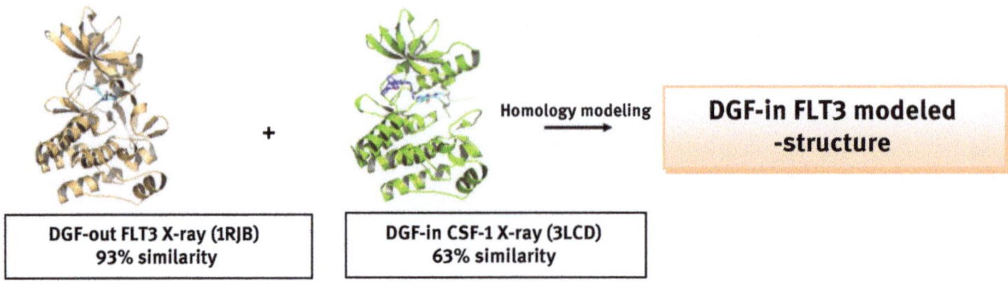

Figure 5.5: Two templates used for model building of the FLT3 structure in the DFG-in conformation.

c-Yes kinase, one of the attractive targets for anti-cancer drugs, adopts active (DFG-in) and inactive (DFG-out) conformations like FLT3, and its inhibitors are classified into type-I and type-II, respectively. A recent work utilized the structural based virtual screening (SBVS) as well as structural based and pharmacophore-based (SB-PB) tandem screening approaches to identify type-I and type-II inhibitors. For the SBVS workflow, the 3D structure of the target was prepared by comparative modeling using human Src kinase with inactive conformation (PDB: 2SRC) as template, followed by simulations for 100 ns in an explicit solvent system. From the 100 ns trajectory, the top seven ensembles were used for the screening process, which resulted in three hits (Z126204226, Z356233398 and Z1338036236). For the SB-PB tandem screening, the active and inactive forms of c-Yes kinase were modeled by employing the active and inactive forms of Ab1 kinase, respectively, as templates and the type-I and type-II inhibitors were identified by screening over both these forms [46].

In another work, 3D structure investigation, binding pattern, dynamic properties and the role of water molecules in the active site of PIM-3 kinase have been reported [47]. The 3D structure of PIM-3 kinase was predicted via comparative modeling, in which a co-crystallized structure of PIM-1 kinase was taken as the template. The reference structure showed a sequence identity of 72% with PIM-3. To understand the binding patterns, molecular docking studies were performed, which revealed the stability of the ligand in the putative binding pocket of the enzyme. Finally, MD simulations revealed that water molecules fill the gap between the ligand and the active site residues by forming hydrogen bonds; as a result, the ligand is stable in the binding pocket.

The human male germ cell associated kinase (hMAK), a target for prostate cancer, has been studied. The CDK2 crystal structure (PDB: 1W8C) was used as a template in the prediction of the 3D structure of the target. The Ramachandran plot of the modeled structure had almost 90% residues in the allowed region, indicating that the built model is reliable. Furthermore, docking studies of five known inhibitors (R547, flavopiridol, AT7519, CHEMBL162, and CHEMBL603469) revealed the competitive nature of the inhibitors. MD simulation trajectories of protein-inhibitor complexes indicated that, except AT7519, all other inhibitors are stabilized throughout the simulations in the protein active site. In addition, the solvated interaction energies, calculated from MD simulation trajectories, revealed the nature of non-bonding contacts and provide valuable insights required for improving the inhibition activity [48].

5.5.1.3 Tubulin

Microtubules are tube-like protein polymers composed of tubulin heterodimers, viz. α and β subunits. They hold prime importance in a number of cellular functions, such as in the development and maintenance of cell shape and structure, cell motility, organelle transport, cell signaling, cell division and mitosis [49]. Numerous studies have highlighted the importance of tubulin isotypes in regulating microtubule dynamics and thus targeting them in order to develop novel anti-cancer agents as a promising method [50, 51].

Kumbhar et al. [52] performed a computational study to predict the origin of the differential binding affinity of DAMA-colchicine for human tubulin isotypes $\alpha\beta_{II}$, $\alpha\beta_{III}$ and $\alpha\beta_{IV}$. The 3D homology models of human αβ tubulin isotypes were prepared taking bovine β_{II} tubulin (PDB: 1SA0) as the template. Their investigation found that DAMA-colchicine prefers the αβ interface of tubulin in 1SA0, $\alpha\beta_{II}$ and $\alpha\beta_{Lv}$. However, a slightly different orientation is observed for $\alpha\beta_{III}$ tubulin. In $\alpha\beta_{II}$ and $\alpha\beta_{IV}$ isotypes, residue Lys350 forms a hydrogen bond with the methoxytropone ring of DAMA-colchicine, whereas, in $\alpha\beta_{III}$, it makes a hydrogen bond with the mercaptoacetyl group of DAMA-colchicine which alters the binding pose of DAMA-colchicine in the binding pocket of $\alpha\beta_{III}$. The overall order of binding energies of the tubulin isotypes for DAMA-colchicine is $\alpha\beta_{IV} \approx \alpha\beta_{II} \quad \alpha\beta_{III}$, which is different from the experimental order.

In the next study, the binding site of three different β-tubulin isotypes, viz. β_I, β_{III} and β_{VI}, were analyzed and the binding modes of various benzimidazole-2-carbametes (BzC) were determined. The 3D structures of the targets were prepared through homology modeling. The D chain of the *Gallus gallus* β-tubulin in complex with nocodazol (PDB: 5CA1) was used as a template for all the three isotypes. The Ramachandran plots for the constructed models showed reliability greater than 98%. The built models were subjected to molecular docking studies, followed by MD simulations. The results showed that the BzC derivatives have higher affinity toward the β_I and β_{III} isotypes. The residues Glu198 and Ser/Cys239 from the β_I and β_{III} isotypes form hydrogen bonds with the the BzC derivatives and hence stabilize them at their binding sites. However, the presence of Ala198 in the β_{VI} isotype reduces the affinity of BzCs [53].

γ-Tubulin, another important isoform of the tubulin family, is an essential component of the microtubule organization center. It plays a crucial role in the nucleation and organization of microtubules during cell division [54, 55], but it has also been found to be over-expressed in many cancer types [56, 57]. Recent studies have highlighted the need of discovering potent γ-tubulin inhibitors that would possibly halt mitosis in cancer cells. In this direction, Suri and Naik [58] studied the binding modes of noscapine, amino-noscapine and bromo-noscapine at the interface of the γ-tubulin dimer. The crystal structure of the γ-tubulin dimer has some missing non-terminal amino acids. These missing amino acids were filled using multiple template homology modeling (PDBs: 3CB2, 2Q1F and 2D2M). All three ligands were docked and the protein-ligand complexes were subjected to simulation for 25 ns. Results revealed that noscapinoids are well accommodated inside the binding cavity and show interaction with both γ-tubulin units at the interface region. Furthermore, binding affinity calculations using MM-PBSA method revealed that bromo-noscapine shows the best binding affinity, followed by noscapine and amino-noscapine. The higher binding affinity of bromo-noscapine was attributed to its tendency to form hydrogen bonds with the residues Met249, Asn250, and Gly247 in the A chain.

5.5.1.4 Proteases

Proteases execute a large variety of functions and have important biotechnological applications [59]. They are currently classified into six broad groups: serine proteases, threonine proteases, cysteine proteases, aspartate proteases, metalloproteases, and glutamic acid proteases.

Cysteine proteases are enzymes containing a specific cysteine residue that catalyses the breaking of peptide bonds. These proteases share a common catalytic mechanism, where the nucleophilic cysteine thiol is a part of the catalytic triad or dyad. These enzymes are found to play numerous indispensable roles, ranging from general catabolic functions, protein processing, and parasite immunoevasion, and have thus emerged as attractive targets in designing effective drugs [60]. Cathepsin B is a cysteine protease which catalyzes the degradation of Aβ (1–42) peptides that are the main constituents of amyloid plaques, causing Alzheimer's diseases [61]. The C-terminal truncation of Aβ (1–42) peptides by cathepsin B reduces the amyloid beta peptide deposition. Activating this enzyme is an effective strategy to treat Alzheimer disease and, therefore, understanding the catalytic mechanism at the atomic level can help in developing activators for this enzyme.

In one study, cysteine protease from the bacteria *Xanthomonascampestris* was studied computationally [62]. The 3D structure of the enzyme was modeled using MODELLER 9v7 and SWISS MODEL. The amino acid sequence of the cysteine protease was retrieved from the NCBI protein sequence database (Accession no.—ZP_06488281.1). Then the BLAST program was used to search for a suitable template in PDB, and human Cathepsin B (PDB: 2IPP) was found to show good similarity. The Ramachandran plot showed that 98% of the residues lie in the favored region, whereas the *Z*-score confirmed that the quality of the model is as good as the X-ray structure. On superimposing the modeled protein with the template protein, the residues Cys17, His87, Gln88 were found to form the active site, and the Aβ peptide [10YDVHHNKLVFF20] (1AML.pdb) was docked into it. It was found that the –SH group of the active site residue Cys17 of the enzyme forms a hydrogen bond with the backbone carboxyl oxygen atoms of Lys16 and Leu17 of the Aβ peptide. These interactions were further confirmed by simulating the protein using MD for 20 ns.

In another study, falcipain-3, a cysteine protease from *Plasmodium falciparum*, was modeled computationally. This enzyme is one of the cysteine proteases, another is falcipain-2, which is implicated in hemoglobin degradation, and hence both are potential targets for designing anti-malarial drugs. A homology model for falcipain-3 was derived based on MSA of the falcipain-3 sequence with the homologs using the COMPOSER module of SYBYL 6.7. At this point, a theoretically reasonable model was constructed with a few trivial abnormalities, which was further corrected by minimization using the DISCOVER module of InsightII. Finally the binding modes of the modeled protein were justified using molecular docking studies. Some differences in the residues lining the S2 pockets of these enzymes were observed, leading to a narrower S2 pocket in falcipain-3, as compared to falcipain-2 [63].

5.5.2 Membrane proteins

5.5.2.1 G protein-coupled receptors

The G protein-coupled receptors (GPCRs) are integral membrane proteins consisting of seven trans-membrane α-helices that are connected by three extracellular and three intracellular loops of variable lengths [64]. GPCRs are involved in information transfer (signal transduction) from outside the cell to the cellular interior and mediate multiple physiological processes [65–68]. Literature survey revealed that GPCRs are responsible for more than 30 different human diseases [69] and almost 50% of the clinically prescribed drugs have been found to target GPCRs [70, 71]. Herein, we report modeling of various important membrane proteins and their inhibitors.

The 5-hydroxytryptamine or serotonin 5-HT_7 receptor, a GPCRs family member, plays a major role in cognition in the central nervous system [72]. In the past, a number of agonist and antagonist based ligands were developed that specifically and selectively target 5-HT_7 receptors [73, 74]. In one of the recent studies, multiple template homology modeling, followed by various filtration processes, was employed to predict the best 3D model of the 5-HT_7 receptor. The workflow designed for the purpose is shown in Figure 5.6. Initially 17 models were generated using mono and multiple templates. The templates used were 2RH1, 1F88 and 3D4S. Results from the initial screening processes revealed that the models based on the template 2RH1 and 3D4S have better stereochemical quality. It was also found that

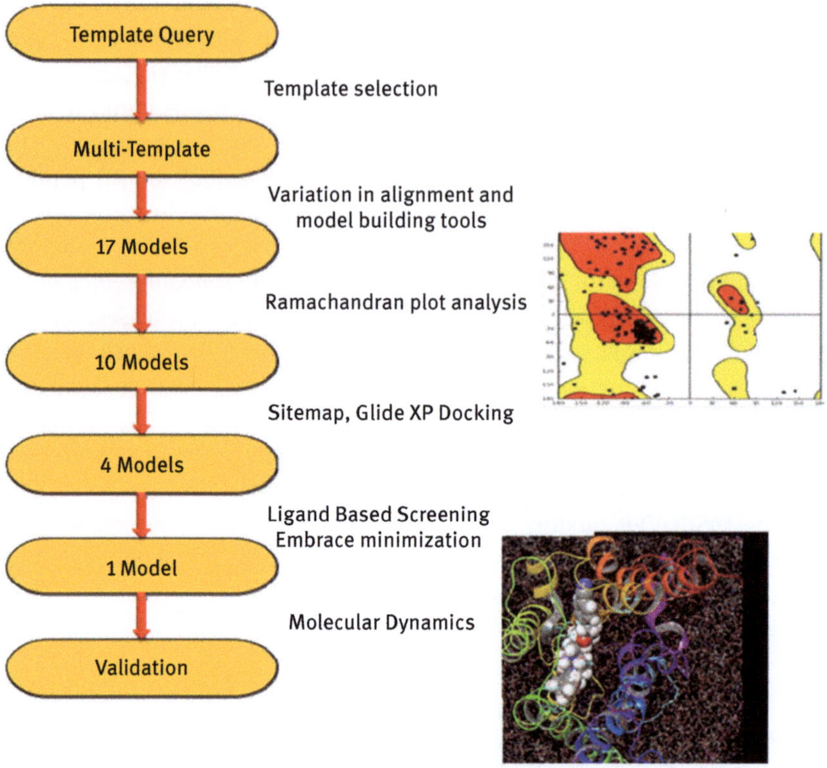

Figure 5.6: The workflow designed for developing the best 3D model for the serotonin 5-HT_7 receptors. It involves selecting the best model out of the 17 models generated using various techniques.

the models built using the mono template approach are superior in quality. Furthermore, docking based screening study with the antagonist ligands of these models showed that only three models possess good binding affinities. These three models were further evaluated and, finally, based on ligand based screening, one model (based on 3D4S) was finally selected. This model was further subjected to MD simulation studies to understand the binding modes of the agonist and antagonist. The results obtained revealed that the residues involved in binding of agonist and antagonist are unique. In addition, the agonist upon binding brings conformational changes that result in stabilization of the overall protein in the agonist bound complex [75].

Other important members of GPCRs are CC chemokine receptors 4, abbreviated as CCR4. They play a crucial role in the progression of asthma, T-cell lymphoma, inflammation, and Alzheimer disease [76–78]. Therefore, CCR4 is a promising drug target, and various studies on its inhibition by small molecules have been reported. Gadhe and Kim [79] reported a computational study involving homology modeling, docking and MD simulation to understand the binding modes in CCR4. The modeled structure for CCR4 was built using the X-ray crystal structure of human CCR5 (PDB: 4MBS) as the template. The built structure had almost 99.3% residues in the most favored and additionally allowed regions of the Ramachandran plot, while approximate 0.7% residues were in the generously allowed region, and no residue was present in the disallowed region, indicating that the model is of good quality. MD simulation studies of the docked complexes, viz. protein-inhibitor, identified the residues involved in ligand binding. Results showed that the interacting residues identified by the docking were displaced, and new residues were found near the ligands during the simulations. Furthermore, Principal Component Analysis revealed that CCR4 unfolds at the ligand binding site and thus some new residues, which may play crucial role in ligand binding, were identified.

Human gonadotropin hormone receptor (GnRH) is the target of several medications used in fertility disorders. Sakhteman et al. [80] built a model for GnRH using seven templates (4N6H, 2RH1, 1GZM, 41AR, 4PHU, 4DJH, and 4GRV) based on ten different threading methods in the modeling process. The modeled receptor had more than 90% of the residues in the favored region, which indicates that the conformational characteristics of the modeled receptor are similar to the native proteins. The prepared model was then subjected to 100 ns MD simulation to sample the different conformations of the receptor. Results revealed that the system attained a steady state after 80 ns and was well equilibrated during the simulation. Large conformational changes were observed for some of the residues located in the extracellular domain and in the cytoplasmic domain of the receptor. This provided several frames representing the receptor at different states. These frames were subjected to cross-docking simulation with some known antagonists to predict the most favorable model of the receptor at the antagonist state.

The human dopamine D4 receptor belongs to the family of dopamine receptors and is a GPCR target for the treatment of neurological and psychiatric conditions [81]. Khoddami and coworkers [18] built a model for the human dopamine D4 receptor using homology modeling and MD simulations. The model for the receptor was predicted using four templates (3PB1, 3UON, 4GRV, and 4IB4). The templates were selected based on their sequence similarity with the D4 receptor sequence. The Ramachandran plot of the model had more than 96% of residues in the favored or allowed region, indicating that the model is reliable. The most conformational fluctuation during the simulation process was observed for some of the residues located in the first extracellular and third cytoplasmic domains. The ionic lock, a common feature of most class A GPCRs, was also found to be formed between the cytoplasmic region of helices III (Arg133) and IV (Glu389) (Figure 5.7). The principal component analysis gave different frames of the receptors by clustering the MD trajectories. Furthermore, docking studies of some known antagonists, together with a

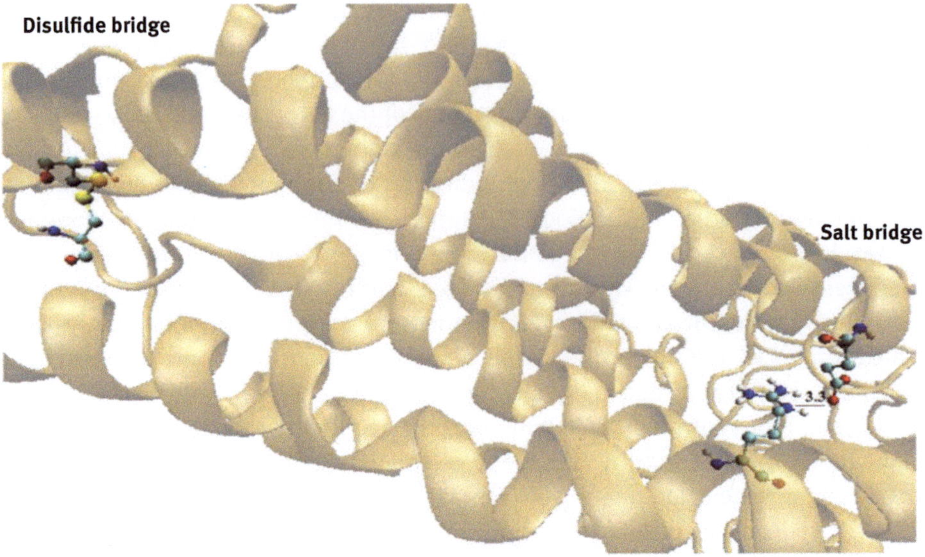

Figure 5.7: Presence of ionic lock and disulfide-bridge showing the two structural features of the receptor. (Reprinted with permission).

series of decoys over these frames (two frames from each cluster), yielded a model that effectively discriminates between active ligands and inactive decoys.

5.5.2.2 P-glycoprotein

P-glycoprotein (P-gp, ABCB1 or MDR1) is a cell membrane ATP-binding cassette (ABC) that continues to be the main focus of research interest. The protein acts as a protector of normal tissues against xenobiotics and has a great impact on the adsorption, distribution, metabolism and elimination properties of a variety of drugs [82]. It is involved in the transport of several chemical and structurally unrelated drug agents and contributes to the multidrug resistance (MDR) in half of human cancers that results in chemotherapeutic failure [83, 84]. There are reports which show that inhibition of P-gp efflux can reverse MDR of cancers [85, 86]. The lack of the crystal structure of human P-gp has limited the rational drug design strategies. It is found to exist in two main conformations: inward-facing and outward-facing, whereas till date only one crystal structure for human P-gp (outward-facing) is available [87]. However, in recent years, structures for related proteins which can serve as templates in the generation of the 3D structure for human P-gp have started to appear. Human P-gp is generally composed of two trans-membrane domains (TMDs) and two nucleotide binding domains (NBDs). Each TMD contains six helices and a large hydrophobic drug binding site that can effectively bind two to three molecules simultaneously.

The homology model for human P-gp has been modeled in three different catalytic states: inward open (IO), intermediate open (IIO) and outward open (OO) using crystal structures of *C. elegans* P-gp [88], human mitochondrial P-gp and Sav1866 P-gp, respectively. A multi-targeted molecular dynamics (MTMD) study has been carried out to investigate the P-gp efflux mechanism and conformational changes at the drug binding site. The IO state was taken as the initial structure for MTMD, whereas the IIO and OO states were considered as the targeted structures. MTMD results revealed that NBDs, which were initially far apart, gradually approached each other and finally dimerized in the OO state, forming a catalytic dyad. Furthermore, significant difference was observed in binding of the substrate and inhibitor. The substrate binds more strongly in the IO state and its binding gradually gets destabilized as the structure changes from the IO to the OO state, whereas the inhibitor remains in stable interactions with the drug binding residues throughout the process [89].

In another study, the 3D structure for human P-gp was modeled using the X-ray structure of Apo murine P-gp (PDB: 3G5U) [90]. The Ramachandran plot of the modeled structure showed that 83.3% of the residues lie in the most favored region, while 0.8% residues are in the disallowed region. However, after the model was refined using MD simulations, 85.4% residues were found in the most favored region and only 0.4% were in the disallowed region. Various dihydropyrimidines derivatives were docked and the studies revealed that the active and inactive compounds showed different types of binding modes with the protein. The active compounds were projected towards the charged polar residues of the active site, namely Arg905, Gln912, and Arg547. Furthermore, it was found that the presence or absence of the nitro group in these derivatives also has a significant effect on their inhibitory activity. In addition, MD studies of the three most active ligands also concluded that the position of the nitro group plays a crucial role in determining the activity, because it forms hydrogen bonding with some important residues like Arg905, Ser909, Ser474, and Val472.

In another study, two piperine analogs were designed and synthesized for their ability to inhibit human P-gp *in silico* and *in vitro*. The modeled structure for the protein showed almost 92.5% residues in the most favored region and no residues in the disallowed regions. On superimposition, the RMSD between the modeled structure and template *C. elegans* (PDB: 4F4C) was found to be 2.68 Å, indicating that the model shows significant structural similarity with the template, viz. *C. elegans*. Molecular docking studies of these two analogs showed the formation of hydrophobic and π-π interactions with certain residues in the active site. MD simulation studies of the docked protein-ligand complexes and apo form of protein have also been reported. Results revealed that an equilibrium condition was achieved within 200 ps of the simulation period. Throughout the simulation process, the backbone of all the protein-ligand complexes was found to be highly stable in comparison to the apo protein, and the hydrophobic interactions between the ligands and drug binding residues were also found to be maintained [91].

5.5.3 Metalloproteins

5.5.3.1 Metallo-endoproteases

Metalloproteins are enzymes that contain a metal ion in the active site for the enzymatic activity. Metallo-endoproteases, metal containing proteases, have been found to be involved in various biological processes, especially in diseases such as cancer and arthritis [92]. Among these, marine alkaline protease (MP), which is an important zinc containing endoproteases, has potential application as a detergent additive. In a study, Ji and coworkers [93] performed comparative molecular modeling studies for the cold-adaptive form of MP. The cold adaptive form of the enzyme is conformationally more flexible, and hence more than one template was used for modeling. The crystal structures of *Pseudomonas aeruginosa* protease, psychrophilic alkaline protease, and *Serratiamarcescens* protease were used as templates and the well validated 3D structure was modeled using MODELLER. Further, MD simulations over a time period of 2 ns with the AMBER11 package clearly showed the occurrence of two flap motions, which provides such flexibility. In the docking studies, Zn^{2+} was found to be bound to the side-chains of His186, His190, His196, Tyr226 and a water molecule.

5.5.3.2 Histone deacetlyases

Histone deacetlyase (HDAC) are a family of proteins involved in the deacetylation of histones and other non-histone substrates. They have been linked to causing cancer and neurological diseases [94, 95]. The HDAC family is formed by 18 members, which are grouped into four classes: class I (1–3 and 8); class II, which is divided into IIa (4, 5, 7 and 9) and IIb (HDAC6 and 10); class III (sirtuins SIR1-7) and class IV (11). Classes I, II, and IV have Zn^{2+} as a cofactor, whereas class III contains nicotinamide adenine dinucleotide (NAD+) and is referred to as sirtuins. Therefore, modeling of each HDAC protein needs to be done specifically for designing selective inhibitors.

The isoform HDAC6 modulates the acetylation of α-tubulin and participates in the microtubule network [96]. There are several domains of this isoform that are crucial for different catalytic activities [97]. Out of these domains, DD1 (87–404 residues), called HDAC-DD1, is the most important and, therefore, its 3D model was developed to study the protein and its inhibitors. This catalytic domain was modeled using the I-TASSER server [98] by exploiting the solved crystal structures of some HDAC proteins having structural identities (2VCG (37%), 1ZZO (37%), 1C3P (29%), 3MEN (36%), 3COY (47%), 3MAX (29%), and 2VQJ (48%)). It is a zinc containing enzyme, so in order to place Zn^{2+} in the catalytically correct position, the authors fed coordinating information from four PBD codes: 1W22, 1T69, 2V5W and 3FOR into the model, assessed the model quality using PROCHECK. Homology modeling showed that the catalytic domain of DD2-HDAC6 has a significantly wider channel rim than HDAC1. After that, 100 ns MD simulation was performed to stabilize the model and, by using the clustering analysis technique using the last 70 ns MD simulation ensembles, the most populated conformation of the protein was obtained. Finally, docking studies were performed on the most commonly occurring form of the protein using the five experimentally reported HDAC6 inhibitors. Hence, in this way a reliable molecular study of this enzyme was carried out [99].

Another important isoform, HDAC10, has been studied recently, where the X-Ray crystal structure of *Daniorerio* (zebrafish) HDAC10 (PDB ID; 5TD7) was used as a template. MODELLER and BIOVIA DS 4.5 were used together for model building and the best model (with lowest normalized DOPE score) was chosen. In order to find novel inhibitors to HDAC10, a few known inhibitors of HDAC 10 were taken from the CHEMBL database and the best binding ligand was chosen. The best binder was then taken forward for ligand-based virtual screening, where the resembling ligands from the ZINC database were filtered. The ligand ZINC19749069 was found to be the most probable inhibitor and MD simulation studies using the complex showed stable binding with this compound [100].

HDACs have been used to treat parasitic infections also, and recently the first X-ray crystal structure of a parasite, *Schistosomamansoni* HDAC8, has been reported. Using this as a template, Melesina and coworkers [101] prepared three dimensional models of *Schistosomahaematobium*, *S. japonicum*, *Clonorchissinensis*, *Echinococcusmultilocularis*, *E. granulosus*, *Taeniasolium* and *Hymenolepismicrostoma* HDAC8. Keeping in view the structural complexity of HDACs, only the catalytic domain of the above parasites was modeled using MODELLER. MD simulations were carried out for the refinement of the models using the AMBER 12 program, where the AMBER ff99SB forcefield27 was employed for the protein and the general Amber force field for the ligand. The force field parameters reported by Hoops et al. [102] were applied for the zinc ion. Low RMSD values over 10 ns suggested a stable structure. Furthermore, molecular docking of known SmHDAC8 inhibitors to all the generated protein models was done to compare the putative binding modes. This study thus paved the path of designing inhibitors which are specific for parasitic HDACs over human HDACs.

5.6 Conclusions

In this work, we have discussed in detail about the combined application of homology modeling, MD, and molecular docking for studying various proteins, such as membrane proteins, globular proteins, and metal containing proteins. We have seen that specific workflows have been developed in each case. For instance, more than one template has been used during homology modeling for conformational active proteins, or those having different catalytic states. Similarly, MD simulations have been performed differently for membrane proteins and metalloproteins. Although there are limitations associated with each technique, such as quality and reliability of the model in homology modeling and high computational cost in MD, a careful and reasonable approach toward these techniques can definitely help in solving various biological problems, which are otherwise not possible to solve experimentally.

Acknowledgements: The authors Varun and Sonam acknowledge financial assistance in the form of Junior Research Fellowship from Council of Scientific and Industrial Research (CSIR) and Senior Research Fellowship (SRF) from University Grants Commission (UGC), respectively.

References

[1] Arora R, Issar U, Kakkar R. Identification of novel urease inhibitors: pharmacophore modeling, virtual screening and molecular docking studies. J Biomol Struct Dyn. 2018;1–5.
[2] Badhani B, Kakkar R. In silico studies on potential MCF-7 inhibitors: a combination of pharmacophore and 3D-QSAR modeling, virtual screening, molecular docking, and pharmacokinetic analysis. J Biomol Struct Dyn. 2017;35:1950–67.
[3] Issar U, Arora R, Kumari T, Kakkar R. Combined pharmacophore-guided 3D-QSAR, molecular docking, and virtual screening on bis-benzimidazoles and ter-benzimidazoles as DNA–topoisomerase I poisons. Struct Chem. 2019;30:1–7.
[4] Spyrakis F, Cavasotto CN. Open challenges in structure-based virtual screening: receptor modeling, target flexibility consideration and active site water molecules description. Arch Biochem Biophys. 2015;583:105–19.
[5] Cavasotto CN, Phatak SS. Homology modeling in drug discovery: current trends and applications. Drug Discov Today. 2009;14:676–83.
[6] Krieger E, Nabuurs SB, Vriend G. Homology modeling. Methods Biochem Anal. 2003;44:509–24.
[7] Meier A, Söding J. Automatic prediction of protein 3D structures by probabilistic multi-template homology modeling. PLoS Comput Biol. 2015;11:e1004343.
[8] Vyas VK, Ukawala RD, Ghate M, Chintha C. Homology modeling a fast tool for drug discovery: current perspectives. Indian J Pharm Sci. 2012;74:1.
[9] Martí-Renom MA, Stuart AC, Fiser A, Sánchez R, Melo F, Šali A. Comparative protein structure modeling of genes and genomes. Annu Rev Biophys Biomol Struct. 2000;29:291–325.
[10] Cheng J. A multi-template combination algorithm for protein comparative modeling. BMC Struct Biol. 2008;8:18.
[11] Liu T, Guerquin M, Samudrala R. Improving the accuracy of template-based predictions by mixing and matching between initial models. BMC Struct Biol. 2008;8:24.
[12] Chatzou M, Magis C, Chang JM, Kemena C, Bussotti G, Erb I, et al. Multiple sequence alignment modeling: methods and applications. Brief Bioinform. 2015;17:1009–23.
[13] Laskowski RA, MacArthur MW, Moss DS, Thornton JM. PROCHECK: a program to check the stereochemical quality of protein structures. J Appl Crystallogr. 1993;26:283–91.
[14] Sippl MJ. Recognition of errors in three-dimensional structures of proteins. Proteins Struct Funct Bioinf. 1993;17:355–62.

[15] Krissinel E, Henrick K. Secondary-structure matching (SSM), a new tool for fast protein structure alignment in three dimensions. Acta Crystallogr D Biol Crystallogr. 2004;60:2256–68.
[16] Pettersen EF, Goddard TD, Huang CC, Couch GS, Greenblatt DM, Meng EC, et al. UCSF Chimera—a visualization system for exploratory research and analysis. J Comput Chem. 2004;25:1605–12.
[17] Feng Z, Pearce LV, Xu X, Yang X, Yang P, Bumberg PM, et al. Structural insight into tetrameric hTRPV1 from homology modeling, molecular docking, molecular dynamics simulation, virtual screening, and bioassay validations. J Chem Inf Model. 2015;55:572–88.
[18] Khoddami M, Nadri H, Moradi A, Sakhteman A. Homology modeling, molecular dynamic simulation, and docking based binding site analysis of human dopamine (D4) receptor. J Mol Model. 2015;21:36.
[19] González MA. Force fields and molecular dynamics simulations. Écolethématique de la SociétéFrançaise de la Neutronique. 2011;12:169–200.
[20] Snow CD, Sorin EJ, Rhee YM, Pande VS. How well can simulation predict protein folding kinetics and thermodynamics? Annu Rev Biophys Biomol Struct. 2005;34:43–69.
[21] Kandt C, Ash WL, Tieleman DP. Setting up and running molecular dynamics simulations of membrane proteins. Methods. 2007;41:475–88.
[22] Sperotto MM, May S, Baumgaertner A. Modelling of proteins in membranes. Chem Phys Lipids. 2006;141:2–9.
[23] Lindahl E, Sansom MS. Membrane proteins: molecular dynamics simulations. Curr Opin Struct Biol. 2008;18:425–31.
[24] Wassenaar TA, Ingólfsson HI, Böckmann RA, Tieleman DP, Marrink SJ. Computational lipidomics with insane: a versatile tool for generating custom membranes for molecular simulations. J Chem Theory Comput. 2015;11:2144–55.
[25] Tian W, Chen C, Lei X, Zhao J, Liang J. CASTp 3.0: computed atlas of surface topography of proteins. Nucleic Acids Res. 2018;46:363–7.
[26] Arora R, Issar U, Kakkar R. In Silico study of the active site of Helicobacter pylori urease and its inhibition by hydroxamic acids. J Mol Graph Model. 2018;83:64–73.
[27] Morris GM, Huey R, Olson AJ. Using autodock for ligand-receptor docking. Curr Protoc Bioinf. 2008;24:8–14.
[28] Jain AN. Surflex: fully automatic flexible molecular docking using a molecular similarity-based search engine. J Med Chem. 2003;46:499–511.
[29] Cui YL, Zhang JL, Zheng QC, Niu RJ, Xu Y, Zhang HX, et al. Structural and dynamic basis of human cytochrome P450 7B1: a survey of substrate selectivity and major active site access channels. Chem Eur J. 2013;19:549–57.
[30] Nelson DR. The cytochrome p450 homepage. Hum Genomics. 2009;4:59.
[31] Dozio E, Briganti S, Vianello E, Dogliotti G, Barassi A, Malavazos AE, et al. Epicardial adipose tissue inflammation is related to vitamin D deficiency in patients affected by coronary artery disease. Nutr Metab Cardiovasc Dis. 2015;25:267–73.
[32] Mauf S, Penna-Martinez M, Jentzsch T, Ackermann H, Henrich D, Radeke HH, et al. Immunomodulatory effects of 25-hydroxyvitamin D3 on monocytic cell differentiation and influence of vitamin D3 polymorphisms in type 1 diabetes. J Steroid Biochem Mol Biol. 2015;147:17–23.
[33] Slominski AT, Brożyna AA, Skobowiat C, Zmijewski MA, Kim TK, Janjetovic Z, et al. On the role of classical and novel forms of vitamin D in melanoma progression and management. J Steroid Biochem Mol Biol. 2018;177:159–70.
[34] Jayaraj JM, Krishnasamy G, Lee JK, Muthusamy K. In silico identification and screening of CYP24A1 inhibitors: 3D QSAR pharmacophore mapping and molecular dynamics analysis. J Biomol Struct Dyn. 2018;37:1700–1714.

[35] Fan JR, Zheng QC, Cui YL, Li WK, Zhang HX. Investigation of ligand selectivity in CYP3A7 by molecular dynamics simulations. J Biomol Struct Dyn. 2015;33:2360–7.

[36] Mordvinov VA, Shilov AG, Pakharukova MY. Anthelmintic activity of cytochrome P450 inhibitors miconazole and clotrimazole: in-vitro effect on the liver fluke Opisthorchisfelineus. Int J Antimicrob Agents. 2017;50:97–100.

[37] Pakharukova MY, Mordvinov VA. The liver fluke Opisthorchisfelineus: biology, epidemiology and carcinogenic potential. Trans R Soc Trop Med Hyg. 2016;110:28–36.

[38] Shukla R, Chetri PB, Sonkar A, Pakharukova MY, Mordivnov VA, Tripathi T. Identification of novel natural inhibitors of Opisthorchisfelineus cytochrome P450 using structure-based screening and molecular dynamic simulation. J Biomol Struct Dyn. 2018;36:3541–56.

[39] Fabbro D, Cowan-Jacob SW, Moebitz H. Ten things you should know about protein kinases: IUPHAR review 14. Br J Pharmacol. 2015;172:2675–700.

[40] Rutaganira FU, Fowler ML, McPhail JA, Gelman MA, Nguyen K, Xiong A, et al. Design and structural characterization of potent and selective inhibitors of phosphatidylinositol 4 kinase IIIβ. J Med Chem. 2016;59:1830–9.

[41] Calmettes G, Ribalet B, John S, Korge P, Ping P, Weiss JN. Hexokinases and cardioprotection. J Mol Cell Cardiol. 2015;78:107–15.

[42] Assouline S, Cocolakis E, Borden K. The development of novel therapies for the treatment of Acute Myeloid Leukemia (AML). Cancers. 2012;4:1161–79.

[43] Vijayan RS, He P, Modi V, Duong-Ly KC, Ma H, Peterson JR, et al. Conformational analysis of the DFG-out kinase motif and biochemical profiling of structurally validated type II inhibitors. J Med Chem. 2014;58:466–79.

[44] Leung AY, Man CH, Kwong YL. FLT3 inhibition: a moving and evolving target in acute myeloid leukaemia. Leukemia. 2013;27:260.

[45] Ke YY, Singh VK, Coumar MS, Hsu YC, Wang WC, Song JS, et al. Homology modeling of DFG-in FMS-like tyrosine kinase 3 (FLT3) and structure-based virtual screening for inhibitor identification. Sci Rep. 2015;5:11702.

[46] Ramakrishnan C, Mary Thangakani A, Velmurugan D, Anantha Krishnan D, Sekijima M, Akiyama Y, et al. Identification of type I and type II inhibitors of c-Yes kinase using in silico and experimental techniques. J Biomol Struct Dyn. 2018;36:1566–76.

[47] Ul-Haq Z, Gul S, Usmani S, Wadood A, Khan W. Binding site identification and role of permanent water molecule of PIM-3 kinase: A molecular dynamics study. J Mol Graph Model. 2015;62:276–82.

[48] Tanneeru K, Balla AR, Guruprasad L. In silico 3D structure modeling and inhibitor binding studies of human male germ cell-associated kinase. J Biomol Struct Dyn. 2015;33:1710–9.

[49] Jordan MA, Wilson L. Microtubules as a target for anticancer drugs. Nat Rev Cancer. 2004;4:253.

[50] Ravanbakhsh S, Gajewski M, Greiner R, Tuszynski JA. Determination of the optimal tubulin isotype target as a method for the development of individualized cancer chemotherapy. Theor Biol Med. 2013;10:29.

[51] Wang Y, Zhang H, Gigant B, Yu Y, Wu Y, Chen X, et al. Structures of a diverse set of colchicine binding site inhibitors in complex with tubulin provide a rationale for drug discovery. Febs J. 2016;283:102–11.

[52] Kumbhar BV, Borogaon A, Panda D, Kunwar A. Exploring the origin of differential binding affinities of human tubulin isotypes αβII, αβIII and αβIV for DAMA-colchicine using homology modelling, molecular docking and molecular dynamics simulations. PLoS One. 2016;11:e0156048.

[53] Aguayo-Ortiz R, Cano-González L, Castillo R, Hernández-Campos A, Dominguez L. Structure-based approaches for the design of benzimidazole-2-carbamate derivatives as tubulin polymerization inhibitors. Chem Biol Drug Des. 2017;90:40–51.
[54] Joshi HC, Palacios MJ, McNamara L, Cleveland DW. γ-Tubulin is a centrosomal protein required for cell cycle-dependent microtubule nucleation. Nature. 1992;356:80.
[55] Oakley BR, Oakley CE, Yoon Y, Jung MK. γ-Tubulin is a component of the spindle pole body that is essential for microtubule function in Aspergillusnidulans. Cell. 1990;61:1289–301.
[56] Katsetos CD, Dráberová E, Legido A, Dráber P. Tubulin targets in the pathobiology and therapy of glioblastomamultiforme. II. γ-tubulin. J Cell Physiol. 2009;221:514–20.
[57] Niu Y, Liu T, Tse GM, Sun B, Niu R, Li HM, et al. Increased expression of centrosomal α, γ-tubulin in atypical ductal hyperplasia and carcinoma of the breast. Cancer Sci. 2009;100:580–7.
[58] Suri C, Naik PK. Combined molecular dynamics and continuum solvent approaches (MM-PBSA/GBSA) to predict noscapinoid binding to γ-tubulin dimer. SAR QSAR Environ Res. 2015;26:507–19.
[59] Engles L. Review and application of serine protease inhibition in coronary artery bypass graft surgery. Am J Health Syst Pharm. 2005;62:S9–14.
[60] Siklos M, BenAissa M, Thatcher GR. Cysteine proteases as therapeutic targets: does selectivity matter? A systematic review of calpain and cathepsin inhibitors. Acta Pharm Sin B. 2015;5:506–19.
[61] Tanzi RE, Bertram L. Twenty years of the Alzheimer's disease amyloid hypothesis: a genetic perspective. Cell. 2005;120:545–55.
[62] Dhanavade MJ, Jalkute CB, Barage SH, Sonawane KD. Homology modeling, molecular docking and MD simulation studies to investigate role of cysteine protease from Xanthomonascampestris in degradation of Aβ peptide. Comput Biol Med. 2013;43:2063–70.
[63] Sabnis YA, Desai PV, Rosenthal PJ, Avery MA. Probing the structure of falcipain-3, a cysteine protease from Plasmodium falciparum: comparative protein modeling and docking studies. Protein Sci. 2003;12:501–9.
[64] Lappano R, Maggiolini M. G protein-coupled receptors: novel targets for drug discovery in cancer. Nat Rev Drug Discov. 2011;10:47.
[65] Chattopadhyay A. GPCRs: Lipid-Dependent Membrane Receptors That Act as Drug Targets. Adv Biol. 2014;2014:1–12.
[66] Fanelli F, De Benedetti PG. Update 1 of: computational modeling approaches to structure–function analysis of G protein-coupled receptors. Chem Rev. 2011;12:438–535.
[67] Klabunde T, Hessler G. Drug design strategies for targeting G-protein-coupled receptors. Chembiochem. 2002;3:928–44.
[68] Rosenbaum DM, Rasmussen SG, Kobilka BK. The structure and function of G-protein-coupled receptors. Nature. 2009;459:356.
[69] Schöneberg T, Schulz A, Biebermann H, Hermsdorf T, Römpler H, Sangkuhl K. Mutant G-protein-coupled receptors as a cause of human diseases. Pharmacol Ther. 2004;104:173–206.
[70] Overington JP, Al-Lazikani B, Hopkins AL. How many drug targets are there? Nat Rev Drug Discov. 2006;5:993.
[71] Schlyer S, Horuk R. I want a new drug: G-protein-coupled receptors in drug development. Drug Discov Today. 2006;11:481–93.
[72] Buhot MC. Serotonin receptors in cognitive behaviors. Curr Opin Neurobiol. 1997;7:243–54.
[73] Demirkaya K, ÖM A, Şenel B, Öncel Torun Z, Seyrek M, Lacivita E, et al. Selective 5-HT7 receptor agonists LP 44 and LP 211 elicit an analgesic effect on formalin-induced orofacial pain in mice. J Appl Oral Sci. 2016;24:218–22.

[74] Medina RA, Sallander J, Benhamú B, Porras E, Campillo M, Pardo L, et al. Synthesis of new serotonin 5-HT7 receptor ligands. Determinants of 5-HT7/5-HT1A receptor selectivity. J Med Chem. 2009;52:2384–92.
[75] Jha P, Chaturvedi S, Swastika , Pal S, Jain N, Mishra AK. Improvising 5-HT7R homology model for design of high affinity ligands: model validation with docking, embrace minimization, MM-GBSA, and molecular dynamic simulations. J Biomol Struct Dyn. 2018;36:2475–94.
[76] Pease JE, Horuk R. Chemokine receptor antagonists: part 1. Expert Opin Ther Pat. 2009;19:39–58.
[77] Yoshie O, Fujisawa R, Nakayama T, Harasawa H, Tago H, Izawa D, et al. Frequent expression of CCR4 in adult T-cell leukemia and human T-cell leukemia virus type 1–transformed T cells. Blood. 2002;99:1505–11.
[78] Yuan Q, Bromley SK, Means TK, Jones KJ, Hayashi F, Bhan AK, et al. CCR4-dependent regulatory T cell function in inflammatory bowel disease. J Exp Med. 2007;204:1327–34.
[79] Gadhe CG, Kim MH. Insights into the binding modes of CC chemokine receptor 4 (CCR4) inhibitors: a combined approach involving homology modelling, docking, and molecular dynamics simulation studies. Mol Bio Syst. 2015;11:618–34.
[80] Sakhteman A, Khoddami M, Negahdaripour M, Mehdizadeh A, Tatar M, Ghasemi Y. Exploring 3D structure of human gonadotropin hormone receptor at antagonist state using homology modeling, molecular dynamic simulation, and cross-docking studies. J Mol Model. 2016;22:225.
[81] Beaulieu JM, Gainetdinov RR. The physiology, signaling, and pharmacology of dopamine receptors. Pharmacol Rev. 2011;63:182–217.
[82] Glavinas H, Krajcsi P, Cserepes J, Sarkadi B. The role of ABC transporters in drug resistance, metabolism and toxicity. Curr Drug Deliv. 2004;1:27–42.
[83] Baumert C, Günthel M, Krawczyk S, Hemmer M, Wersig T, Langner A, et al. Development of small-molecule P-gp inhibitors of the N-benzyl 1, 4-dihydropyridine type: novel aspects in SAR and bioanalytical evaluation of multidrug resistance (MDR) reversal properties. Bioorg Med Chem. 2013;21:166–77.
[84] Szakács G, Paterson JK, Ludwig JA, Booth-Genthe C, Gottesman MM. Targeting multidrug resistance in cancer. Nat Rev Drug Discov. 2006;5:219.
[85] Shukla S, Ohnuma S, Ambudkar SV. Improving cancer chemotherapy with modulators of ABC drug transporters. Curr Drug Targets. 2011;12:621–30.
[86] Thomas H, Coley HM. Overcoming multidrug resistance in cancer: an update on the clinical strategy of inhibiting p-glycoprotein. Cancer Control. 2003;10:159–65.
[87] Kim Y, Chen J. Molecular structure of human P-glycoprotein in the ATP-bound, outward-facing conformation. Science. 2018;359:915–9.
[88] Prajapati R, Singh U, Patil A, Khomane KS, Bagul P, Bansal AK, et al. In silico model for P-glycoprotein substrate prediction: insights from molecular dynamics and in vitro studies. J Comput Aided Mol Des. 2013;27:347–63.
[89] Prajapati R, Sangamwar AT. Translocation mechanism of P-glycoprotein and conformational changes occurring at drug-binding site: insights from multi-targeted molecular dynamics. Biochim Biophys Acta Biomembr. 2014;1838:2882–98.
[90] Shahraki O, Zargari F, Edraki N, Khoshneviszadeh M, Firuzi O, Miri R. Molecular dynamics simulation and molecular docking studies of 1, 4-Dihydropyridines as P-glycoprotein's allosteric inhibitors. J Biomol Struct Dyn. 2018;36:112–25.
[91] Syed SB, Arya H, Fu IH, Yeh TK, Periyasamy L, Hsieh HP, et al. Targeting P-glycoprotein: investigation of piperine analogs for overcoming drug resistance in cancer. Sci Rep. 2017;7:7972.

[92] Coussens LM, Fingleton B, Matrisian LM. Matrix metalloproteinase inhibitors and cancer—trials and tribulations. Science. 2002;295:2387–92.

[93] Ji X, Wang W, Zheng Y, Hao J, Sun M. Homology modeling and molecular dynamics simulation studies of a marine alkaline protease. Bioinform Biol Insights. 2012;6:BBI-S10663.

[94] Kashyap K, Kakkar R. An insight into selective and potent inhibition of histone deacetylase 8 through induced-fit docking, pharmacophore modeling and QSAR studies. J Biomol Struct Dyn. 2019;1–22.

[95] Witt O, Deubzer HE, Milde T, Oehme I. HDAC family: what are the cancer relevant targets? Cancer Lett. 2009;277:8–21.

[96] Tran AD, Marmo TP, Salam AA, Che S, Finkelstein E, Kabarriti R, et al. HDAC6 deacetylation of tubulin modulates dynamics of cellular adhesions. J Cell Sci. 2007;120:1469–79.

[97] Zhang Y, Gilquin B, Khochbin S, Matthias P. Two catalytic domains are required for protein deacetylation. J Biol Chem. 2006;281:2401–4.

[98] Zhang Y. I-TASSER server for protein 3D structure prediction. BMC Bioinf. 2008;9:40.

[99] Sixto-López Y, Bello M, Rodríguez-Fonseca RA, Rosales-Hernández MC, Martínez-Archundia M, Gómez-Vidal JA, et al. Searching the conformational complexity and binding properties of HDAC6 through docking and molecular dynamic simulations. J Biomol Struct Dyn. 2017;35:2794–814.

[100] Ibrahim Uba A, Yelekçi K. Homology modeling of human histone deacetylase 10 and design of potential selective inhibitors. J Biomol Struct Dyn. 2019;37:3627–36.

[101] Melesina J, Robaa D, Pierce RJ, Romier C, Sippl W. Homology modeling of parasite histone deacetylases to guide the structure-based design of selective inhibitors. J Mol Graph Model. 2015;62:342–61.

[102] Hoops SC, Anderson KW, Merz KM, Jr. Force field design for metalloproteins. J Am Chem Soc. 1991;113:8262–70.

Janvhi Machhar, Ansh Mittal, Surendra Agrawal, Anil M. Pethe and Prashant S. Kharkar

6 Computational prediction of toxicity of small organic molecules: state-of-the-art

Abstract: The field of computational prediction of various toxicity end-points has evolved over last two decades significantly. Availability of newer modelling techniques, powerful computational resources and good-quality data have made it possible to generate reliable predictions for new chemical entities, impurities, chemicals, natural products and a lot of other substances. The field is still undergoing metamorphosis to take into account molecular complexities underlying toxicity end-points such as teratogenicity, mutagenicity, carcinogenicity, etc. Expansion of the applicability domain of these predictive models into areas other than life sciences, such as environmental and materials sciences have received a great deal of attention from all walks of life, fuelling further development and growth of the field. The present chapter discusses the state-of-the-art computational prediction of toxicity end-points of small organic molecules to balance the trade-off between the molecular complexity and the quality of such predictions, without compromising their immense utility in many fields.

Keywords: predictive toxicology, NCEs, mutagenicity, carcinogenicity, teratogencity, environmental toxicity

6.1 Introduction

Candidate attrition in late-stage clinical development can be very troublesome for the obvious reasons. This is even more bothersome in the "blockbuster era" where every pharmaceutical company dream of developing a blockbuster. The company's very existence, at times, depends on the hopes and ultimate success garnered by so-called "wish-to-be-a-blockbuster" drug. The modern-day drug discovery and development researchers have learnt their lessons during clinical development, mostly in retrospect, post establishment of recombinant DNA (rDNA) technology. In last decade or so, the pharmaceutical productivity in terms of new molecular entities (NMEs) entering in the clinic, has actually gone down, in spite of the availability of the state-of-the-art technologies such as combinatorial

This article has previously been published in the journal Physical Sciences Reviews. Please cite as: Machhar, J., Mittal, A., Agrawal, S., Pethe, A. M., Kharkar, P. S. Computational prediction of toxicity of small organic molecules: state-of-the-art *Physical Sciences Reviews* [Online] 2019, 4 DOI: 10.1515/psr-2019-0009.

https://doi.org/10.1515/9783110631623-006

chemistry, high-throughput screening and many others, at our disposal [1]. The need to adopt drug discovery and development approach fundamentally different from the 'conventional' one was strongly felt by those who were directly or indirectly involved in the 'art of bringing new medicines to the market' [2]. Despite the gloomy picture over the last decade-and-half, the global pharmaceutical R&D convincingly succeeded, particularly in overturning late-stage failures and creating success stories, ensuring themselves that "there was still hope". Thanks to all the early *in vitro* screening assays and the subsequent computational tools based on the data generated by these assays.

Of the several reasons for late-stage clinical failure, poor pharmacokinetics (PK) and toxicity were the front-runners [3]. The term "poor PK" literally meant inadequacy in any property or process directly or indirectly related to "absorption, distribution, metabolism and excretion (ADME)" such as solubility, permeability, metabolic stability, etc., of the investigational drug. The conventional assays for accessing the PK in early-stage drug discovery were either inadequate, low-throughput or were used only after the clinical candidate was nominated. The molecular biology techniques like rDNA technology for producing the biomacromolecules to be used in these *in vitro* assays just started appearing in mid-1980s and were far from realizing their full potential. The researchers evaluated these PK properties then by what was available to them. A large number of methods for evaluating the PK/toxicity end-points were in practice, leading to data quality or homogeneity issues. Moreover, the unavailability of such data in public domain was a major issue hindering development of computational or predictive models for these relevant drug properties. It was only in late-1990s that the pharmaceutical industry started appreciating the usefulness of early ADME/T (T for toxicity) studies to curb the late-stage clinical failures significantly, which actually became a reality, bringing down the earlier attrition-rate due to poor ADME/T from 50% to <10% [4]. The philosophy was simple – "Fail early, fail cheap".

The early identification of problematic candidates relived the stress on the pharmaceutical R&D. Systematic efforts in harmonizing the early-ADME/T (eADME/T) assays and their adoption into medium-to-high throughput made high-quality data available to the modeller community [5]. The after-effect was totally awesome. Several computational or *in silico* models for simple physicochemical to complex ADME/T end-points were reported [6]. Another factor motivating the growth of predictive ADME/T field was, of course, the newer modelling techniques and improved computational power. It was possible to handle and use large volumes of data, mostly from high-throughput eADME/T assays, as input for model development, validation and subsequent predictions. Few online databases, benchmarking datasets and web interfaces appeared on the internet so that scientific community could use them for developing and validating their own predictive models using newer methodologies [6].

In recent times, the pharmaceutical companies and other stakeholders are looking forward to pre-competitive data sharing, potentially unfolding so-called "proprietary" quality data, which can be extremely useful for developing highly predictive models [7, 8]. In addition, several databases such as Human Metabolome Database version 4.0 [9–12], TOXNET [13], PubChem [14] and many others are publicly available. The data has potentially been and being used for developing several predictive models and tools for eADME/T. Overall, most, if not all, of the stakeholders are on the same page for making high-quality computational models available to the drug discovery researchers furthering their significant contribution in improving the pharmaceutical productivity by discovering safer and efficacious drugs.

A large collection of research articles, reviews, perspectives, books, and book chapters are published year-by-year on "predictive ADME/T, computational PK, *in silico* toxicity, computational toxicity, predictive toxicology, predictive biopharmaceutics and PK and predictive PK modeling". Predictive PK also covers related terms such as pharmacokinetic-pharmacodynamic (PK/PD) modelling and population PK. Recent published texts and articles cover the predictive PK and related fields [15–20]. The reader is encouraged to refer to these reading materials for gaining an in-depth view of the topic. No further discussion on the predictive PK is included in this chapter.

By and large, toxicity properties outnumber the PK properties or associated end-points. Historically, the toxicity end-points have been difficult to model, mostly due to their complex nature, experimental determination and potential impact on the overall drug discovery and development. During early-stage drug discovery, a large array of studies is conducted in preclinical animal models to demonstrate safety for further first-in-human use, i. e. phase I clinical trials. Post-thalidomide disaster back in early 1960s, the Food and Drug Administration (FDA) amended drug safety requirements from time to time which has made the drug development process really cumbersome. The literature is full of stories including drug recalls and withdrawals due to safety concerns. Most notable example being voluntary withdrawal of rofecoxib (Vioxx®), a selective cyclooxygenase 2 inhibitor due to cardiovascular toxicity related to its mechanism of action [21]. The unethical abstinence from disclosure of extremely important toxicity issues during preclinical and developmental stages by the sponsors can be very problematic for the patients and the regulators. In brief, thorough knowledge on toxicity end-points along the drug development pathway is extremely essential, which of course, does not guarantee clinical success.

Evaluation of mutagenicity, genotoxicity, carcinogenicity, teratogenicity, hepatotoxicity, nephrotoxicity, acute systemic toxicity, developmental toxicity, reproductive toxicity, cardiovascular toxicity, CNS toxicity, repeat-dose toxicity and chronic toxicity are critical for the subsequent investigational new drug application, requesting FDA to allow phase I studies. In addition, the environmental toxicity or ecotoxicity has gained much attention in the recent past. Historically, the main emphasis has been on developing potent, selective, and metabolically-stable drugs with minimal toxicity to major organs and tissues. Due to the very nature of small organic compounds, they tend to interact with several off-targets precipitating large number of side-effects or adverse reactions. At times during post-marketing surveillance (phase IV) stage, idiosyncratic reactions may lead to drug withdrawal. These and similar facts necessitate a thorough understanding of potential toxicity issues with an NME. The scientific community has devoted significant amounts of resources to develop predictive toxicity models over the years. Recent literature (in terms of absolute numbers) only emphasizes the importance of this field (Figure 6.1). The computational models for ADME/T end-points, reported prior to 2016, have been extensively discussed elsewhere along with freely-available web tools and commercial softwares [22]a. The current book chapter narrates the recent developments (majorly beyond 2015) in the *in silico* or computational toxicity prediction of small organic molecules such as drugs, drug-like compounds, drug impurities, degradants, natural products, chemical intermediates, speciality chemicals, environmentally benign and hazardous chemicals and other related substances. Relatively bigger organic molecules (>1 kD) were not covered since the predictive power of these models wears off beyond this molecular weight cut-off, i. e. with increasing molecular complexity. Several reported, publicly available and commercial *in silico* models have been discussed and their predictive performance, applicability domain, and

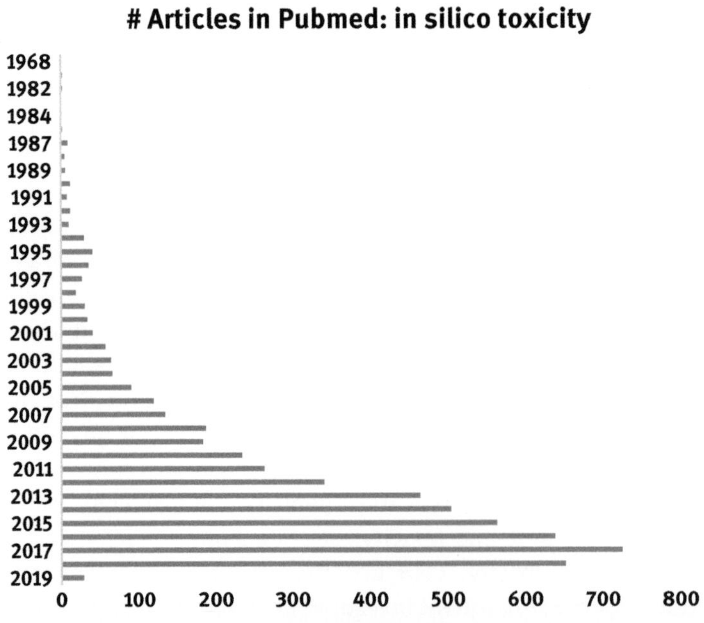

Figure 6.1: Year-wise distribution of articles listed in PubMed using term "in silico toxicity" (1968–2019).

limitations highlighted so as to give a beginner or a seasoned researcher an in-depth overview of the predictive/computational toxicology field. Since the models "improvise" as and when newer data is available, model development and refinement is a continuous process. An eye on the latest developments in the field is more than necessary. The present book chapter is a step in this direction.

6.2 Computational models for various toxicity end-points

6.2.1 Genotoxicity, mutagenicity and carcinogenicity

Genotoxicity involves damage to the genetic material (DNA, chromosomes, etc.), irrespective of the mechanism by which it is induced, whereas mutagenicity involves direct interaction of a chemical agent with DNA, possibly leading to a mutation with subsequent carcinogenesis (process by which a chemical induces benign/malignant tumours), if any. The DNA-reactive substances may cause damage to a single gene, gene segment, gene block or chromosome, even at lower concentration, giving rise to mutations and carcinogenesis. These substances may be drugs, impurities, degradants or other chemicals. Controlling the levels of such substances is an important aspect of drug safety. The International Conference on Harmonisation (ICH) of Technical Requirements for Registration of Pharmaceuticals for Human Use established M7 guidelines for detecting and controlling DNA reactive impurities in pharmaceuticals and to limit their carcinogenesis risks [23]. The efforts on the global scale reinforce the importance of these mutagenic substances; so is the need for computational models which predict these end-points with great accuracy. Of these models, quantitative structure-(toxicity)property relationship (QS(T)PR) models are simple to derive and most of the times, easy to interpret. The key to understanding the potential (geno)toxicity (or mutagenicity) of a query structure is the ability of the model or the tool based on the model to identify the structural alerts for that (geno)toxicity end-point. This simplifies the whole interpretation part of the model. The QS(T)PR models for understanding chemically induced mutagenicity have been extensively discussed [22]b. On similar lines, *in silico* methods for carcinogenicity assessment have been thoroughly explained by Golbamaki et al. [22]c. Readers will benefit greatly after reading these comprehensive texts.

Speaking of structural alerts, certain structural classes such as primary aromatic amines (pAAs), *N*-nitroso, aflatoxin-like and alkyl-azoxy compounds form so-called "cohort of concern". These mutagenic functional groups have theoretically high potential for carcinogenic risk. A gold standard for experimentally assessing mutagenicity risk for chemicals and correlating mutagenicity with animal carcinogenicity, is bacterial reverse mutation assay or Ames test, discovered back in 1970. Given the cost and time involved in the Ames test, it has to be used sparsely during early drug discovery. Hence, computational tools based on Ames data, are used widely for mutagenicity prediction. Given the inherent unpredictable characteristics of the test, select few classes of compounds such as pAAs, despite being potentially mutagenic, are spared and identified as non-mutagenic. This is possibly due to the very fact that the pAAs are converted to their *N*-hydroxylamine (phase I/functionalization reaction) form, and subsequently to acetate, sulfonate or glucuronide conjugates (phase II/ conjugation reaction). These metabolites are highly mutagenic. This is obviously risky, threatening drug attrition, of a candidate originally identified as non-mutagen, may be in the later stages of clinical development.

In an interesting study, Patel et al. attempted to address and discuss this very issue for pAAs so that their mutagenicity potential could be predicted accurately [24]. The efforts by a precompetitive collaborative group for sharing data and the resulting consortium for the Investigation of Genotoxicity of Aromatic Amines were aimed at collecting, publishing and using the proprietary summary Ames data for literature aromatic amines to avoid redundant testing in Ames test and of course, improve the computational model accuracy. A carefully curated set of 268 compounds (189 Ames-negative and 79 Ames-positive – unbalanced dataset) was divided into four structural subclasses (SC1–SC4) and subjected to mutagenicity prediction by four tools. The outcome was mixed predictions. There was no clear trend in the predictions, although they showed good negative predictivity. Despite this observation, the "inconclusive" and "out-of-domain" predictions add the compounds to the load, i. e. assays to run, for experimental Ames test. A fructitious brain-storming in the consortium members resulted in thorough, mechanism-based understanding of the discrepancies between the false-positives and false-negatives. One of the possible solutions to this problem was consensus scoring. But this would leave out those compounds which were predicted as "inconclusive" or "out-of-domain" by either of the models used in the scoring. Another solution could be an expert review of the "incorrect" predictions. At times, the route cause for such discrepancy could be in the structure itself, i. e. the effect of neighbouring groups on the supposed structural alert, in this case, the pAA. Conclusively, we need more and more data in such cases, as it becomes available, just so we improve the prediction power of the model (rather a set of models) moving forward.

Glück et al. evaluated genotoxic and carcinogenic potential of 609 phytochemicals (secondary plant metabolites) present in food using computational models [25] implemented in several tools such as (a) Virtual models for property Evaluation of chemicals within a Global Architecture (VEGA) platform version 1.1.1 [26] containing four models each for mutagenicity and carcinogenicity; (b) LAZAR (lazy structure–activity relationship) [27] models, each one for rodents (multiple species/sites) for carcinogenicity and *Salmonella typhimurium* for mutagenicity predictions and (c) T.E.S.T. (Toxicity Estimation Software Tool) version 4.2 [28] developed at the United States Environmental Protection Agency for mutagenicity predictions. Another software tool – Organisation for Economic Co-operation and Development (OECD) QSAR Toolbox [29] was used for collecting experimental carcinogenicity and mutagenicity data for the query compounds (#609) mentioned above. The models from one tool and the combined models from more than one tools were further evaluated and compared using accuracy, sensitivity, specificity and the Youden's index (sensitivity + specificity – 1). The meticulous modelling efforts led to intriguing outcomes such as (a) the combined models performed better than the model(s) from single tool; (b) average scores of 0.66 and 0.33 defined the upper and lower bounds, respectively, for classification of a compound as mutagenic (carcinogenic) (score > 0.66), non-mutagenic (non-carcinogenic) (score < 0.33) or inconclusive (score between 0.33 and 0.66). The outcome of the study was to identify phytochemicals from different foods as potentially mutagenic (carcinogenic), in addition to corroborating the usefulness of the combined and/or consensus models in improving the prediction quality of the models. Few of the potentially mutagenic, genotoxic and carcinogenic chemicals (including drugs) are shown in Figure 6.2.

In a related study, Di Sotto et al. investigated the genotoxicity of piperitenone oxide (a.k.a. rotundifolone), a natural flavouring agent using *in vitro* assays such as Ames test, comet assay and micronucleus (MN) assay [30] and compared the experimental outcome with *in silico* predictions using VEGA [26] and Toxtree [31]. The *in vitro* studies confirmed the mutagenic and genotoxic properties which were in agreement with the *in silico* results. This was not surprising owing to the presence of epoxide and α, β-unsaturated functionalities in the monoterpene structure (**11**, Figure 6.2). The authors expressed the need for toxicological libraries of naturally occurring chemicals for further derivation of computational models for their toxicity end-point predictions. In addition, the authors also emphasized the importance of overall bioavailability and metabolic fate of such compounds on their genotoxicity profile. It was the first publication featuring the genotoxicity profiling of the natural flavouring agent, **11**, as claimed by the authors.

Bossa et al. have thoroughly reviewed computational (statistical and expert rule-based) models reported in the literature, for predicting genotoxicity/mutagenicity and carcinogenicity of chemicals, with a view to understanding the scientific rationale and regulatory requirements [32]. The authors have thoroughly outlined the experimental assays for carcinogenicity and genotoxicity evaluation along with the open-source, free and commercial computational tools for predicting them. An in-depth understanding and availability of chemical databases in the public domain is extremely important in developing, validating and fine-tuning the predictive models from time to time. Of particular significance is the central role played by regulations and guidance documents such as (a) EU REACH (European Union Registration, Evaluation, Authorization and Restriction of Chemicals) [33] in stimulating sharing of precompetitive information, use of alternatives to animal testing and non-testing approaches; (b) ICH M7 guidelines for assessing DNA reactive impurities in pharmaceuticals [23] and European Food Safety Authority (Panel on Plant Protection Products and their Residues [34]) approach for assessing genotoxicity of their metabolites. The readers are highly encouraged to refer to this book chapter by Bossa et al. for detailed

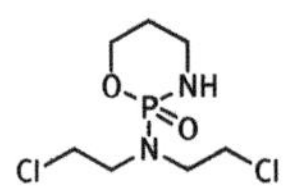
Cyclophosphamide (**1**)

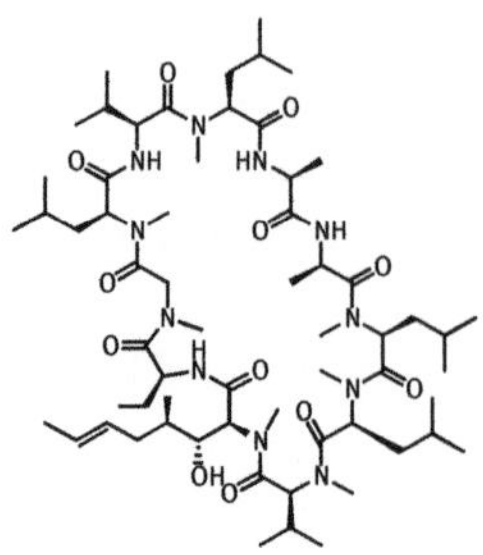
Ciclosporin (**2**)

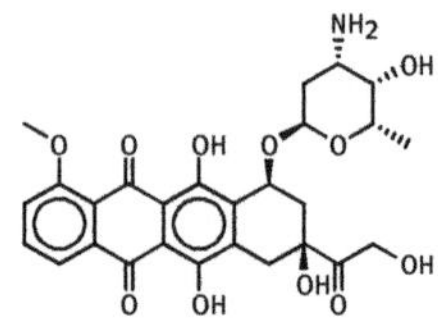
Doxorubicin (**3**)

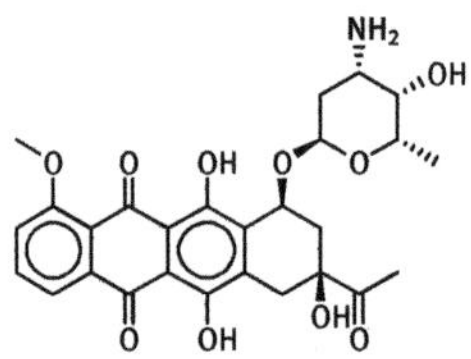
Daunorubicin (**4**)

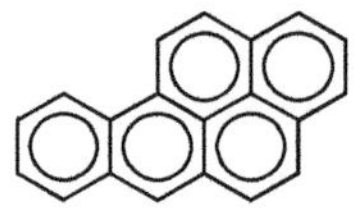
Benzo[a]pyrene (**5**)

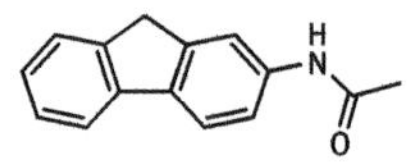
2-Acetylaminofluorene (**6**)

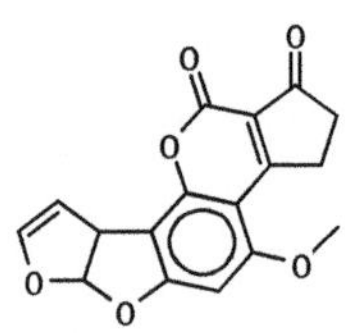
Aflatoxin B1 (**7**)

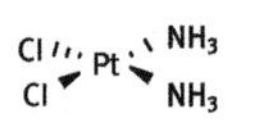
Cisplatin (**8**)

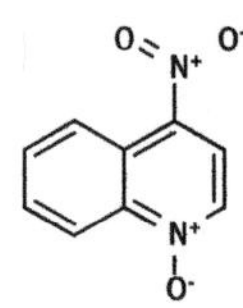
4-Nitroquinoline N-oxide (**9**)

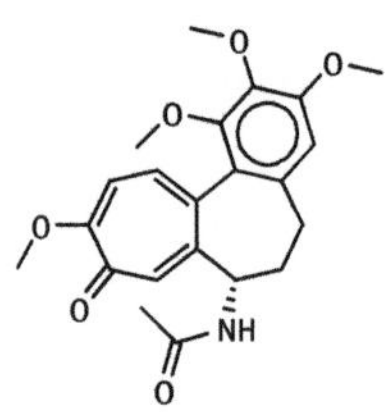
Colchicine (**10**)

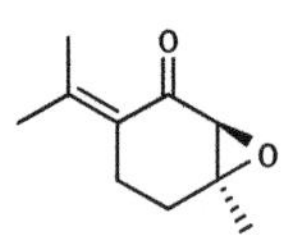
Rotundifolone (**11**)

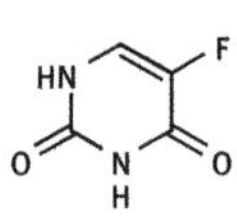
5-Fluorouracil (**12**)

Figure 6.2: Molecular structures of few representative mutagenic, genotoxic and carcinogenic chemicals.

guidance [32], which highlight the worldwide efforts by all the stakeholders in promoting the utility of refined computational models for predicting important toxicity endpoints – genotoxicity and carcinogenicity. In a recent article, Verheyen et al.

thoroughly reviewed the *in vitro* and *in silico* methods of testing mutagenicity potential of chemicals, particularly from regulatory perspective on safety assessment of chemicals [35]. The two types of computational models – expert knowledge- or rule-based and statistical, offer unique advantages and limitations with respect to predictions. The FDA's regulatory draft guidance (2008) on genotoxic and carcinogenic impurities in drug substances and drug products clearly indicated the usefulness of statistical models for identifying structural alerts for known and expected impurities [36].

The toxicity end-point is directly related to the chemical structure of the query chemical. Over the years, problematic substructures or so-called structural alerts or toxicophores for a typical toxicity end-point have been identified. Presence of one or more structural alert(s) in the query structure flags it as potentially toxic chemical. Recently Plošnik et al. published mutagenic and carcinogenic structural alerts and their corresponding mechanism of toxic action in their paper [37]. The ability of these structural alerts, with or without photo- or bioactivation (e. g. *N*-hydroxylation of pAAs), to covalently modify DNA, RNA or histones can be attributed to their mutagenic and carcinogenic potential. Commonly found toxicophores include aromatic nitro groups, alkyl hydrazines, thio- or nitrogen mustards, acyl halides, epoxides, aziridines, quinones, and many others (Figure 6.3). Few of these alerts do not damage the genetic material (non-genotoxic). Rather they induce carcinogenesis by affecting expression of certain genes (epigenetic changes), e. g. aromatic halides, steroidal oestrogens, thiocarbonyls, etc. (Figure 6.3). The most common reactions of this type are DNA (hyper/hypo)methylation and histone acetylation. At times, the mere presence of structural alerts in a molecule does not guarantee toxicity *in vivo*. This could possibly be due to metabolic inactivation or latency owing to various physicochemical and PK properties of the chemicals. Several of the modern toxicity prediction algorithms or tools such as CASE Ultra [38] map these toxicophores back onto the molecular structure (Figure 6.4) so that the chemist could possibly work around those substructures to completely negate their toxicity potential, e. g. lead discovery and optimization of new chemical entities (NCEs). Literature reports on bioisosteric replacement of toxicophores with less hazardous groups are available [39], emphasizing the importance of safety risks posed by the toxicophores and the strategies to circumvent such issues.

The present-day genotoxicity and carcinogenicity tests focus mostly on sensitivity, trading off the specificity and accuracy. To address this very important issue, Fujita et al. resorted to integrated testing strategies (ITSs) incorporating *in vitro* and *in silico* data [40]. They investigated the relationships among genotoxicity test results, carcinogenicity test results and the chemical properties of the test articles using decision tree (optimized decision tree and random forest (RF)) models. A total of 230 database molecules (184 carcinogens + 46 non-carcinogens) from Carcinogenicity Genotoxicity eXperience (CGX) data set, with experimental testing results from Ames, *in vitro* chromosomal aberration, and *in vitro* MN assays

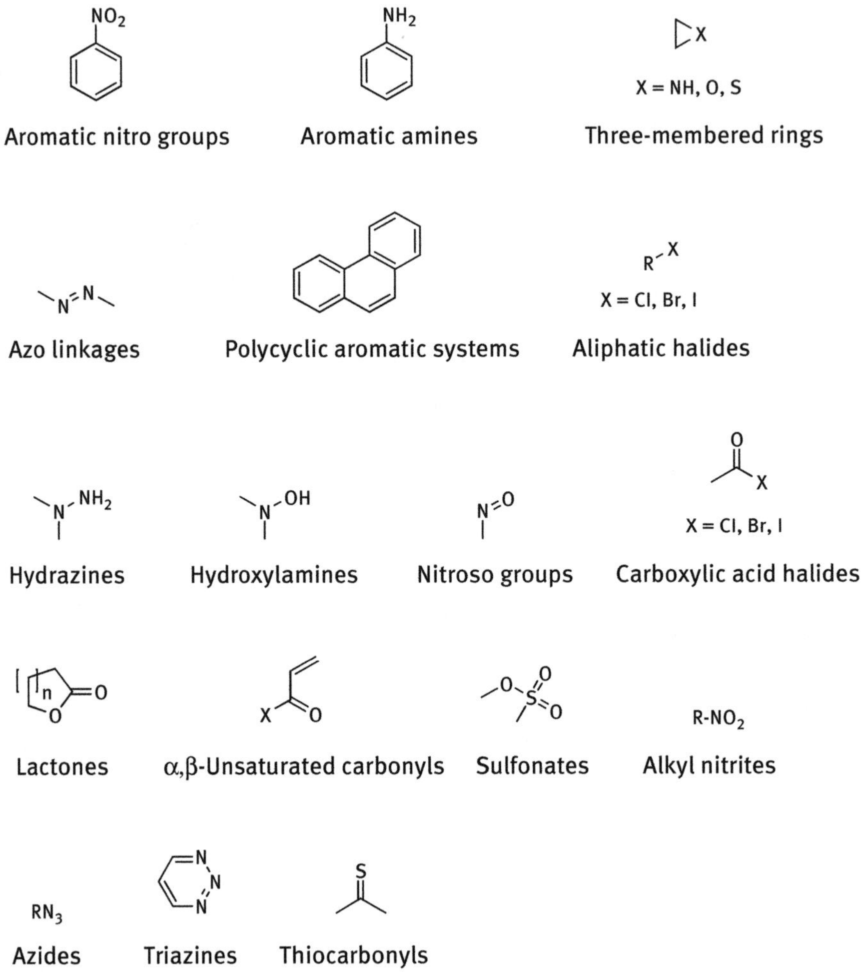

Figure 6.3: Commonly occurring toxicophores in small organic chemicals.

were used in the training set. For analysis of various *in silico* properties for the dataset molecules, QSAR Toolbox version 4.1 [41] was used. Next the genotoxicity test results were compared with the carcinogenicity results. Data pretreatment was done to reduce the bias due to unbalanced sample. In the actual model development, machine learning technique – decision tree – was used. The accuracy was further improved with the help of RF method. The developed models were compared in relation to their performance in predicting their carcinogenicity. The two models exhibited higher cross-validation accuracy (71.5% for optimized decision tree and 75.5% in case of decision forest model) over the regulatory decision tree (54.1%). Further, during the model optimization, troublesome functional

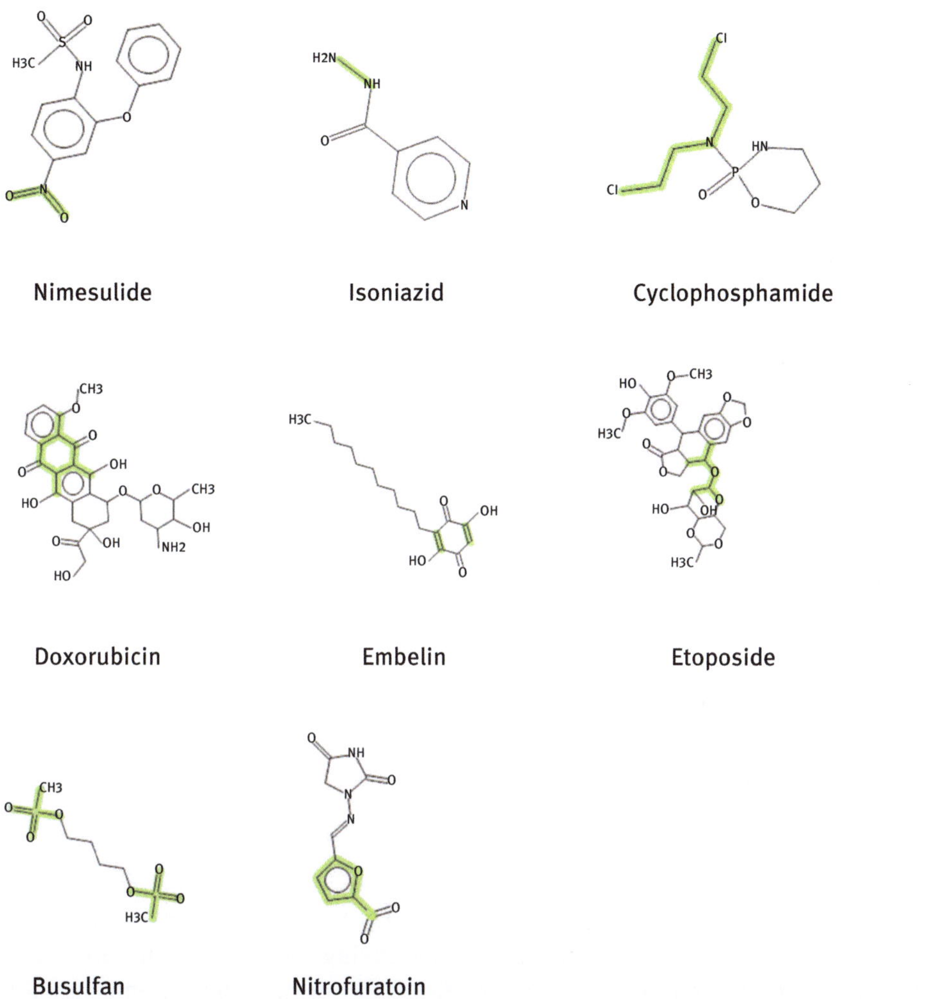

Figure 6.4: Structural alerts mapped back onto the query chemical structure (generated using CASE Ultra predictions for select few compounds.

groups such as amide, thioamide, aryl halide and heteroalicyclic substructures, causing false prediction in genotoxicity and non-genotoxic carcinogenicity, were identified. Most of the models contained *in vitro* genotoxicity parameters, emphasizing their importance in predicting the carcinogenicity. This was claimed to be the first report utilizing ITSs in order to improve the predictive outcome of a toxicity end-point.

In an interesting paper, Fan et al. constructed a dataset of 641 diverse molecules with *in vivo* MN assay results (264 positive or 377 negative), which was further partitioned into training (90% cases) and validation (10% cases) sets [42]. Further processing involved (i) calculation of six molecular fingerprints namely, CDK, CDK Extended, Estate, MACCS, Pubchem and Substructure fingerprints using PaDEL-Descriptor [43], (ii) calculation of molecular descriptors, e. g. constitutional, Basak, Burden, CATS and MOE-type descriptors, using ChemSAR [44] and, (iii) feature selection using standard methods such as intervariable correlation, recursive feature elimination with linear kernel support vector machine (SVM), etc. A total of six machine learning methods were used to build models implemented in Orange 2.0 [45] namely, SVM, naïve Bayes (NB), k-nearest neighbor, C4.5 decision tree, RF and artificial neural network (ANN). The developed models were extensively processed using external diverse validation set. The calculation of accuracy, sensitivity, specificity and receiver operating characteristic (ROC) curve further established the model robustness and predictive power.

The applicability domain was determined using similarity-based measures such as Tanimoto coefficient, leading to compound classification as in-domain and out-of-domain (OD). Additionally, structural alerts (high-frequency fragments in MN-positive chemicals) were identified using SAR in python (SARpy) [46]. A total of 15 models were developed and tested for performance using external validation set wherein all the models exhibited high accuracy (>0.840). MACCS_RF, Pubchem_ANN, Descriptor_RF and Pubchem_SVM (accuracy > 0.9) made to the list of top four. The identified structural alert included aromatic nitro compounds, benzimidazoles, benzidines, anilines, aziridines, epoxypropanes, thiophosphates, nitriles, aromatic diazo compounds and formamides or thioformamides. The most interesting feature of this study was the free accessibility to all the used tools.

The literature contains a large number of models for a given toxicity endpoint, e. g. Ames test. At times, the researchers feel lost as to which model is better than the rest. Another issue is for the scarcity of reliable predictive models for a given toxicity due to variability in the underlying mechanisms and detection methods, e. g. chromosomal damage. Morita et al., in their logical and thought-provoking work, compared the three QSAR models – Derek Nexus, ADMEWorks and CASE Ultra – using CGX dataset to get an in-depth overview of the issue [47]. The dataset contained 440 chemicals (325 carcinogens and 115 non-carcinogens). The best sensitivity was exhibited by CASE Ultra model (91%) over the other two models (Derek 56% and ADMEWorks 67.7%). A similar trend in accuracy of the models was observed. With respect to certain chemical classes, the model performances were equivalent. The models positively predicted well-known chemical classes such as *N*-nitroso, N-nitro, halides, epoxides, alkylating agents, among others. For a similar comparison in case of *in vivo* MN test, CASE Ultra models exhibited higher sensitivity but low specificity. The original issue of poor predictive ability of the chromosomal damage by the predictive models was proposed to be resolved by precise refinement of the training data used for building the model(s).

6.2.2 Developmental toxicity

The toxicity end-point is a complicated one. The OECD guidelines 426 (Developmental Neurotoxicity Study), 416 (Two-Generation Reproduction Toxicity), 415 (One-Generation Reproduction Toxicity), 414 (Prenatal Developmental Toxicity Study), 443 (Extended One-Generation Reproductive Toxicity Study), 421 (Reproduction/Developmental Toxicity Screening Test), 422 (Combined Repeated Dose Toxicity Study with the Reproduction/Developmental Toxicity Screening Test) extensively cover developmental and reproductive toxicity studies, which make up an important cluster of tests to assess a chemical with respect to safety and human health hazards. The detailed description of these tests can be found somewhere else [48]. Overall, these studies are complicated, making the model development process very difficult. The underlying processes involved in reproductive and developmental toxicity include, but not limited to, effects on foetal growth (retardation, decreased weight), foetal survival (pre- or post-implantation loss, death), structural dysmorphogenesis, visceral organ toxicity, neuronal damage, and immunology. The major challenge in model development is the scarcity of reliable input data. All these factors make things very difficult. Marzo et al., in their publication, narrated four *in silico* models including two publicly available models in the VEGA [26] – first being a statistical QSAR classification model named "Computer-Assisted Evaluation of industrial chemical Substances According to Regulations" (CAESAR), and the second being "expert rule-based model" which was adapted from Procter & Gamble (PG) model. The latter two of the four models were part of commercial softwares such as Leadscope Model Applier and CASE Ultra [22]d.

The CAESAR dataset comprised of 201 positives and 91 negatives (total 292), out of which 234 were used as training data and the remaining 58 as test set. The developed statistical model, based on "random forest" implementation, was then used for flagging a chemical as toxicant or non-toxicant. The prediction is based on the comparison of the query molecule with the training data, with respect to molecular descriptors and the resulting similarity. In comparison, the PG data set was larger (total of 716 compounds: 665 positives and 16 negatives and remaining 35 with missing data). The model was based on 641 training data points with developmental toxicity information and composed of 25 categories and six nodes. The "expert rule-based" system classified a compound as toxic if it were found to belong to any of the 25 categories, else non-toxic for developmental toxicity. The commercial models are not discussed here due to the lack of information on the proprietary data used for model development and the higher frequency with which these models are updated. The readers are encouraged to find latest information on these models from the software vendor's websites.

Lu et al. used a large dataset of 17,120 compounds for six toxicity end-points namely, acute toxicity, mutagenicity, tumourigenicity, skin and eye irritation, reproductive effects and multiple dose effects [49]. The dataset was partitioned into training and test sets (4:1) using Kennard-Stone algorithm. Two molecular fingerprints – extended-connectivity fingerprints maximum diameter 4 (ECFP_4) and feature-class fingerprint of diameter 4, molecular descriptors, and chemical-chemical interactions, were used as input variables in the model building process. Of all the generated models, ECFP_4 + LLL proved to be the best during external test set predictions; the balanced accuracy ranged from 0.599 to 0.692. The identified toxicophores (using Laplacian-modified Bayesian model) were mapped back onto the molecular structures, giving a straightforward interpretation of the given models for the six toxicity end-points. The dataset size was large enough to generate models with reasonable prediction power. Further refinement of these models with newer data, as and when available, is likely to improvise on the current version.

To address the time-consuming and expensive assays for evaluating developmental toxicity end-point, Zhang et al. developed an *in silico* model for developmental toxicity using NB classifier method [50]. The authors extensively reviewed the earlier reports on computational prediction of this toxicity using QSAR and expert rule-based models. A unique dataset of 290 drugs containing FDA-defined classes A and B (non-toxicants) and C, D and X (potential developmental toxicants) was used. Data partitioning was done to include 232 drugs in the training set and 58 drugs in the test set (8:2 distribution). Further 63 molecular descriptors and extended-connectivity fingerprints maximum diameter 6 (ECFP_6), a type of topological fingerprints, were used as input variables. Genetic algorithm (GA) based method was used for feature selection. The model development process involved generation of a classifier for non-toxicants and toxicants using NB classifier method. The sensitivity, specificity, positive predictive value and negative predictive value of NB classifier model, NB-1, were 87.6%, 94.4%, 97.2%, and 77%, respectively, with overall prediction accuracy of 89.7%. The robustness of the model was evaluated by external test set with similar results. The molecular features such as distribution coefficient (logD), molecular fractional polar solvent-accessible surface area, molecular polar surface area and number of rings were found to be the most relevant molecular descriptors for the toxicity end-point prediction. The final model, NB-2, was comparable to NB-1 in terms of the prediction parameters. The ROC index of the test set was 0.881. Generation of additional models NB-3 and NB-4 precisely established the importance of ECFP_6 fingerprints in predicting developmental toxicity. The identified structural fragments contributing to developmental toxicity included sulfonyl, sulphonamide, secondary amino, amido, chlorophenyl and ketone groups, to name a few. The developed models may not be comparable with earlier literature models due to different training and test data. Nevertheless, the study unequivocally proved the utility of the classifiers in modelling a complicated data and toxicity end-point such as developmental toxicity.

Hessel et al. [51] in their paper emphasized the importance of designing testing methods encompassing the integral mechanistic landscape of an awfully complicated toxicity end-point such as developmental neurotoxicity. The authors proposed ontology approach which would potentially integrate a computational model for predicting this toxicity. The efforts to build the ontology, in ideal scenario, should integrate chemical, biological and toxicological knowledge. This, in turn, is directly relevant to rate-limiting key steps with a binary outcome – 1: the adverse event would occur and 0: the adverse event would not occur. These events are required to be translated into key animal-free assays, with a readout parameter related to compound exposure. Overall, this integrative approach involving ontology-based modelling of human brain development and the associated toxicity pathways is likely to yield relevant information useful for integrating it into potentially useful *in silico* model, predicting the hazards associated with compound exposure.

Overall, the predictive models for developmental toxicity are scarce, due to the complexity behind the molecular pathways. Successful translation of the *in vitro* to *in vivo* to *in silico* data for meaningful prediction of the toxicity end-point is likely to yield obvious benefits in terms of cost and time.

6.2.3 Hepatotoxicity

Liver is an important organ for disposition of nutrients, xenobiotics and other accidentally administered chemicals. The literature is full of models of predicting hepatotoxicity of chemicals. Hewitt et al. have extensively reviewed models for predicting hepatotoxicity reported between 2000 and 2015 [22]e. Of the total 21 models, 15 were statistical models while six were expert rule-based models. Readers are encouraged to refer to the text for in-depth analysis and related information. López-Massaguer et al. reported a novel approach to hepatotoxicity prediction using systems biology modelling of disturbed metabolic pathways as judged from gene expression data [52].

The workflow was pretty simple and logical involving – (a) collection of gene expression data in human hepatocytes in presence of a chemical from LINCS1000 database; (b) estimation of upper and lower bounds of metabolic reactions; (c) computation of network perturbation by mapping gene expression variability in presence of chemicals and (d) final interpretation due to chemical perturbation. The numbers associated with this study were – 22,119 perturbations for chemicals with gene expression data focusing on 50 most variable (under/overexpressed) genes due to chemical presence and the therapeutically relevant perturbation dose (10 μM). Recon 2, a consensus metabolic model representing human cell metabolism containing cell-type specific models, e. g. liver hepatocytes, was used for flux variability analysis in human hepatocytes. The overall analyses led to identification of altered metabolic pathways under the influence of chemical perturbation, which may be related to the underlying toxicity mechanism. The presented approach based on systems biology is more relevant to *in vivo* situation, strengthening the confidence in the computational models, based on such data. The limiting factor is the availability of such gene expression data for the query chemical, which may be a NCE or some random chemical. A similar report by Carbonell et al. attempted to predict the hepatotoxicity employing systems biology modelling of disturbed metabolic pathways as inferred from gene expression data [53]. Using a tool named flux variability analysis, the authors demonstrated the utility of this systems biology-based novel approach in predicting the hepatotoxicity of three statins – pravastatin, simvastatin and rosuvastatin.

A range of drug toxicity prediction models using machine learning methods such as RF, SVM, näive Bayesian and back propagation neural network, have been thoroughly reviewed by Zhang et al. [54] with the aim of providing state-of-the-art in the predictive toxicology field. The authors highlighted the notable advances in predicting carcinogenicity, mutagenicity and hepatotoxicity. The reviewed models were compared with respect to accuracy, sensitivity and specificity so as to understand and appreciate their applicability and limitations. Banerjee et al. developed ProTox-II, a web server for toxicity prediction of chemicals [55, 56]. It is based on molecular similarity, pharmacophore modelling, fragment behaviour (aka structural alert), and machine learning models predicting hepatotoxicity, acute toxicity, cytotoxicity, mutagenicity, carcinogenicity, and many other relevant toxicity end-points. The best part is that the models are based on data from *in vitro* assays and *in vivo* models. The output of the prediction process is delivered for 33 in-built models with confidence scores along with an overall toxicity radar chart and toxicity targets (Figure 6.5). A simple, yet scientifically elegant, tool like this is likely to generate more interest in the toxicologists, regulatory agencies, modellers, medicinal chemists, pharmacologists, and many others. Continual refinement of such models will only harness their predictive power and of course, the overall utility.

Working on the similar lines, Lu et al. developed predictive models for evaluating hepatotoxicity of drug metabolites using an ensemble approach based on SVM [57]. To address a major toxicity end-point such as drug-induced liver toxicity, the authors constructed QSAR models using 64 hepatotoxic drug metabolites and 3339 non-hepatotoxic drug metabolites from MDL Metabolite database. An intelligent approach was applied to the highly imbalanced dataset by randomly sampling the non-hepatotoxic metabolite population individually to create balanced samples (datasets), generating independent classifiers. Further, all individual classifiers were put together in an ensemble approach with subsequent application of minimum Redundancy Maximum Relevance feature selection method in order to select the molecular descriptors. For the external test set predictions, a Bayesian inference method was used for the toxicity end-point prediction for the metabolites. The resultant model was good with sensitivity 74.14%, specificity 82.77%, and average balanced accuracy 78.47%. The corresponding values for the external predictions were 70%, 65.19% and 60.38%, respectively. The relevant molecular descriptors were related to molecular frontier orbital energy, intramolecular bonding strength and molecular polarity.

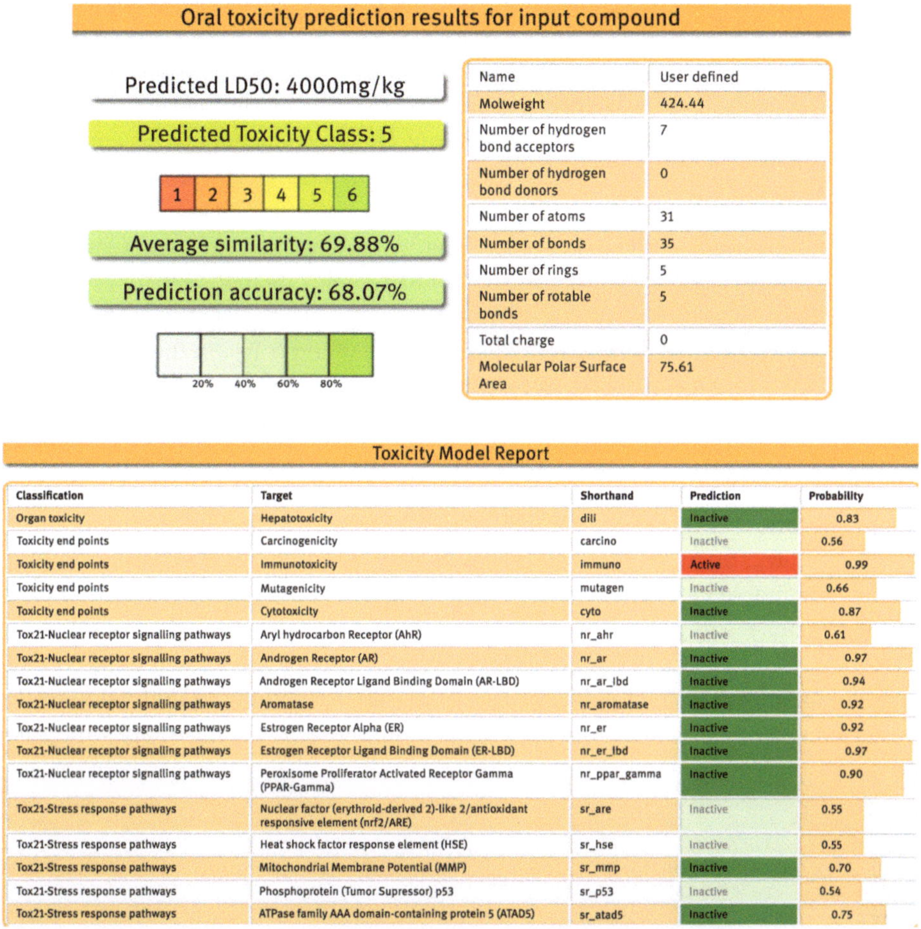

Oral toxicity prediction results for input compound

Predicted LD50: 4000mg/kg

Predicted Toxicity Class: 5

1 2 3 4 5 6

Average similarity: 69.88%

Prediction accuracy: 68.07%

20% 40% 60% 80%

Name	User defined
Molweight	424.44
Number of hydrogen bond acceptors	7
Number of hydrogen bond donors	0
Number of atoms	31
Number of bonds	35
Number of rings	5
Number of rotable bonds	5
Total charge	0
Molecular Polar Surface Area	75.61

Toxicity Model Report

Classification	Target	Shorthand	Prediction	Probability
Organ toxicity	Hepatotoxicity	dili	Inactive	0.83
Toxicity end points	Carcinogenicity	carcino	Inactive	0.56
Toxicity end points	Immunotoxicity	immuno	Active	0.99
Toxicity end points	Mutagenicity	mutagen	Inactive	0.66
Toxicity end points	Cytotoxicity	cyto	Inactive	0.87
Tox21-Nuclear receptor signalling pathways	Aryl hydrocarbon Receptor (AhR)	nr_ahr	Inactive	0.61
Tox21-Nuclear receptor signalling pathways	Androgen Receptor (AR)	nr_ar	Inactive	0.97
Tox21-Nuclear receptor signalling pathways	Androgen Receptor Ligand Binding Domain (AR-LBD)	nr_ar_lbd	Inactive	0.94
Tox21-Nuclear receptor signalling pathways	Aromatase	nr_aromatase	Inactive	0.92
Tox21-Nuclear receptor signalling pathways	Estrogen Receptor Alpha (ER)	nr_er	Inactive	0.92
Tox21-Nuclear receptor signalling pathways	Estrogen Receptor Ligand Binding Domain (ER-LBD)	nr_er_lbd	Inactive	0.97
Tox21-Nuclear receptor signalling pathways	Peroxisome Proliferator Activated Receptor Gamma (PPAR-Gamma)	nr_ppar_gamma	Inactive	0.90
Tox21-Stress response pathways	Nuclear factor (erythroid-derived 2)-like 2/antioxidant responsive element (nrf2/ARE)	sr_are	Inactive	0.55
Tox21-Stress response pathways	Heat shock factor response element (HSE)	sr_hse	Inactive	0.55
Tox21-Stress response pathways	Mitochondrial Membrane Potential (MMP)	sr_mmp	Inactive	0.70
Tox21-Stress response pathways	Phosphoprotein (Tumor Supressor) p53	sr_p53	Inactive	0.54
Tox21-Stress response pathways	ATPase family AAA domain-containing protein 5 (ATAD5)	sr_atad5	Inactive	0.75

Figure 6.5: Sample report from ProTox-II web server tool for a hypothetical molecule.

Zhu et al. reported an *in silico* method predicting drug-induced liver injury (DILI) from adverse drug reaction reports [58] using a carefully collected and curated (from online adverse drug reports) novel dataset containing 122 DILI-positive and 932 DILI-negative drugs. As usual, the unbalanced sample (dataset) posed the obvious problems. To deal with it, the authors used under-sampling the majority class, synthetic minority over-sampling technique and adjusting decision threshold approach as main strategies for developing predictive classification models. Fine-tuning of the RF models (using CDK, MACCS and mold2 descriptors) based on either of the three strategies improved the predictive power. The structural alerts, identified by a graph mining algorithm when applied to hepatotoxic and non-hepatotoxic drugs, such as pyrimidines, purines and halogenated hydrocarbon were found to be crucial for DILI. Despite the age- and gender-dependent reported incidences of

DILI, such a predictive model is only likely to improve our understanding of this toxicity end-point. Going forward, the model may be improvised as and when newer data for varied chemotypes are available.

Yet another critical study focused on computational identification of proteins directly/indirectly involved in DILI using drug-target interaction as the underlying process [59]. Ivanov et al. employed a dataset of 699 drugs (severe 178, moderate 310 and no-DILI 211 drugs) for prediction of interactions with 1534 human proteins. The predictions were subjected to clustering, gene ontology, gene expression and pathway analyses leading to identification of 61 proteins (responsible for disruption of cellular pathways crucial for hepatocyte survival, causing hepatitis, for example) contributing to DILI. The direct utility of this information is lies in the very fact that if the query compound is likely to interact with any or many of these proteins, it will potentially cause DILI. Isn't that amazing? Indeed, it is! Similar computational binary classification model was reported by Toropova et al. [60]. They built predictive models for DILI using Monte Carlo method as implemented in CORAL software [61]. The reasonably predictive classifier was based on the SMILES representation of the molecular structures and H-suppressed graph from the CORAL software. Overall, the simple model demonstrated its usefulness for predicting relatively complicated toxicity end-point. The higher number of reports on *in silico* models for hepatotoxicity emphasizes the growing interest of the modeller community in handling complex toxicity end-points. Such efforts are only going to add value to the field. This is likely to mature the field enough to predict certain toxicity with great accuracy.

6.2.4 Ecotoxicity

The world has become environment conscious in the last decade or so like never before. The global industrialization has polluted the environment, water bodies and soil. This has damaged the ecosystem irreversibly. Consistent rise in population has made matters worse. Right from the time we get up, we are exposed to several products such as personal care products, colorants, foods, pesticides, pharmaceuticals, to name a few. The continued release of these persistent organic compounds in the environment has after-effects, especially due to their non-degradative behaviour. In addition, the improper disposal of the expired pharmaceuticals, the effluents from chemical factories exhibit the toxic effects on the aquatic animals, which on consumption by the human, enter the food-chain, exhibiting deleterious effects on health. The field of QSAR model development for ecotoxicity of organic chemicals (including pharmaceuticals) is not new. Roy et al. extensively reviewed the *in silico* models of ecotoxicity of pharmaceuticals lately [22]f in context with the growing utility of ecotoxicity prediction models due to the environmental damage done by the pharmaceuticals released irresponsibly. The book chapter thoroughly described the applicability of the reported *in silico* models in helping basic risk management and fate in the environment. Kar et al. published an in-depth analysis of risk assessment of the pharmaceuticals in the environment using QSAR modelling approaches [62]. Once in the environment, the released pharmaceuticals undergo degradation and metabolism by the living systems, exploding the possible number of degradants and metabolites. This further necessitates the risk assessment of these additional chemicals originating from the pharmaceuticals. The authors reported an extensive list of ecotoxicity end-points, databases and expert systems used for prediction of ecotoxicity of pharmaceuticals, which can be further extended to chemicals relevant in material sciences.

de Morais e Silva et al. developed a predictive QSAR model-based set of theoretical Volsurf molecular descriptors utilizing the fish acute toxicity values for the low toxicity compounds (Mode of Action 1, MoA 1, non-reactive compounds) [63]. The main objective of this exercise was to dig out molecular properties related to this mechanism. The developed partial least squares model did reasonably well in terms of relevant statistics (internal and external validation). Many physicochemical properties in addition to hydrophobicity, symmetric distribution of the hydrophobic moieties in the molecular structure, and shape, i. e. degree of branching, were found to be important in explaining the variability in the response variable. Overall, the authors demonstrated the utility of such an approach in modelling ecotoxicity end-points and improvise on the predictions which is a direct function of the model quality.

Sangion et al. developed new QSAR models for predicting acute toxicity of contaminants of emerging concern such as active pharmaceutical ingredients in three aquatic organisms representing three main aquatic trophic levels – algae, Daphnia and two species of fish [64]. A software QSARINS was used for MLR-OLS model building using theoretical molecular descriptors calculated by free PaDEL software and selected by GA. After establishing the robustness of the models, acute toxicity prediction for large number of drugs without any experimental data was carried out. Pattern recognition on the predictions was carried out using Principal Component Analysis, named as Aquatic Toxicity Index (ATI). The potential uses of ATI included the ranking of pharmaceuticals or organic chemicals, for that matter, with respect to their toxicity to aquatic environment. Such a model will help medicinal chemists to rule out proposed molecules with potential for aquatic toxicity.

6.2.5 Future perspectives

In addition to the discussed toxicity end-points, attempts were made to search *in silico* models for other toxicity end-points such as renal toxicity (kidney and urinary tract adverse effects in human, e. g. nephropathy, urolithiasis, bladder disorders, etc.), cardiotoxicity (adverse effects on the heart and cardiovascular system, e. g.

human cardiac conduction disorders, human coronary artery disorders, human cardiac failure, human tachycardia, etc.), skin and eye toxicity (irritation and sensitization potential upon contact with skin or eye, e. g. sensory irritation), mammalian reproductive toxicity (reproductive system disorders and adverse effects such as fertility in male/female, sperm toxicity, new-born behavioural toxicity, etc.), teratogenicity (rabbit, mouse, rat, mammal, etc.), acute toxicity models (acute toxicity endpoints such as LD_{50}, MTD, etc.) and ecotoxicity models (bioconcentration factor, biodegradability, toxicity to environmental bacteria, toxicity to fish, etc.). There were hardly any models reported in the literature for most of these end-points. This was not surprising, given the underlying complexity in measuring these toxicities. Also, the data for many of these end-points is not available in the public domain.

Commercial vendors such as MultiCASE license these models developed using FDA data as part of Research Cooperation Agreement [65]. That leaves us with a pertinent question – Who, when and how the models for complicated toxicity end-points such as those listed above, will see the light at the end of tunnel? Such models can only be developed as and when the proprietary data is available to the modelers. Also, the *in silico* models based on the simplified representation of the complicated biological process generating toxic phenotype may serve the intended purpose of prediction of complex toxicities. The *in silico* models may even help in generating data after careful selection of the *in vitro* models from amongst a group of models. The utility of these computational models in both directions, forward and reverse, mainly lies in *in vitro* to *in silico* to *in vivo* correlation, wherever applicable. The onus, in this case, lies with FDA to open the access to their databases to those who are formally "interested" on the non-exclusive basis. The modeller community is eagerly looking forward to such quality and relevant data. In addition, the consensus model, based on a set of models possibly covering mutually exclusive aspects of various individual processes defining the gross phenotype, i. e. toxicity end-point, e. g. developmental neurotoxicity.

The most critical requirements then include quality data, willingness to share pre-competitive and/or competitive data without breaching data integrity, newer modelling techniques and the involvement of domain experts – toxicologists, biologists, chemists, regulators, academicians, software developers, modellers and many others – each contributing in one or the other significant ways for making the whole exercise extremely useful for the betterment of human, animal and environmental health.

6.3 Conclusions

The willingness to contribute to the field of *in silico* prediction of toxicity end-points by all the stakeholders has brought this very field from its nascent stage to relatively matured stage. The challenges are manifold; the foremost being the availability of quality data. Over the years, the mind set of key researchers from corporate world has changed. Access to precompetitive data, establishment of consortia to address common issues, collaborative efforts exchanging expertise, benchmarking datasets for comparison of newer methods, heightened scale of modelling operations and many such activities have fuelled the growth of this field. Modelling complex end-points require a thought process right from the usable data generation, which, in turn, involves selection of proper *in vitro* model with meaningful readout, representing the final end-point. The *in silico* models are facilitating collaborative efforts by the biologists and the modellers to achieve a common goal of developing predictive models. The work flow does not stop here. The real essence is to continually improve the model quality and ultimately the predictive ability of the developed models. In short, it is a continuous and iterative process. The extreme obsession to achieve perfection in the predictability of the futuristic models in the years to come will, no doubt, help in maturing this very important discipline. Several things are at stake, but the willingness and efforts are not!

Acknowledgements: The authors are extremely thankful to the Editor, Prof. Ponnadurai Ramasami, University of Mauritius, Mauritius, for his patience during preparation of this Chapter. PK is thankful to MultiCASE for giving the trial license of CASE Ultra and META Ultra for few months during the calendar year 2018. PK, AP and SA are thankful to Dr Bala Prabhakar, Dean, School of Pharmacy and Technology Management, SVKM's NMIMS, for her constant support and encouragement.

Conflict of Interest: The authors declare no conflict of interest.

References

[1] Paul SM, Mytelka DS, Dunwiddie DT, Persinger CC, Munos BH, Lindborg SR, et al. How to improve R&D productivity: the pharmaceutical industry's grand challenge. Nat Rev Drug Discov. 2010;9:203–14.
[2] Kennedy T. Managing the drug discovery/development interface. Drug Discov Today. 1997;2:436–44.
[3] Prentis RA, Lis Y, Walker SR. Pharmaceutical innovation by the seven UK-owned pharmaceutical companies (1964–1985). Br J Clin Pharmacol. 1988;25:387–96.
[4] Caldwell GW. Compound optimization in early- and late-phase drug discovery: acceptable pharmacokinetic properties utilizing combined physicochemical, in vitro and in vivo screens. Curr Opin Drug Discov. 2000;3:30–41.
[5] Sinko PJ. Drug selection in early drug development: screening for acceptable pharmacokinetic properties using combined in vitro and computational approaches. Curr Opin Drug Discov. 1999;2:42–8.
[6] Kharkar PS. In silico absorption, distribution, metabolism and excretion. In: Markus L, editor. In silico drug discovery and delivery. London: Future Science Group, 2013:149–62.
[7] Removing barriers to data sharing in pharmaceutical research and development. https://www.pharmaceutical-technology.com/features/removing-barriers-data-sharing-pharmaceutical-rd/. Accessed: 13 Jan 2019.

[8] This will change pharma as we know it – what will fuel innovation in healthcare? https://www.elsevier.com/connect/this-will-change-pharma-as-we-know-it-what-will-fuel-innovation-in-healthcare. Accessed: 13 Jan 2019.
[9] Wishart DS, Tzur D, Knox C, Eisner R, Guo AC, Young N, et al. HMDB: the human metabolome database. Nucleic Acids Res. 2007;35:D521–6.
[10] Wishart DS, Knox C, Guo AC, Eisner R, Young N, Gautam B, et al. HMDB: A knowledgebase for the human metabolome. Nucleic Acids Res. 2009;37:D603–10.
[11] Wishart DS, Jewison T, Guo AC, Wilson M, Knox C, Liu Y, et al. HMDB 3.0 — the human metabolome database in 2013. Nucleic Acids Res. 2013;41:D801–7.
[12] Wishart DS, Feunang YD, Marcu A, Guo AC, Liang K, Vázquez-Fresno R, et al. HMDB 4.0 — the human metabolome database for 2018. Nucleic Acids Res. 2018;46:D608–617.
[13] TOXNET: the toxicology data network. https://toxnet.nlm.nih.gov/. Accessed: 13 Jan 2019.
[14] Kim S, Chen J, Cheng T, Gindulyte A, He J, He S, et al. PubChem 2019 update: improved access to chemical data. Nucleic Acids Res. 2019;47:D1102–9.
[15] Benet LZ. Predicting pharmacokinetics/pharmacodynamics in the individual patient: separating reality from hype. J Clin Pharmacol. 2018;58:979–89.
[16] Segall M. Integrating metabolism and toxicity properties. In: Wilson AGE, editor. RSC drug discovery series. new horizons in predictive drug metabolism and pharmacokinetics. London: Royal Society of Chemistry, 2015:227–44.
[17] Yang Y, Zhao Y, Duan JZ, Zhao P, Zhao L, Zhang X. Predictive biopharmaceutics and pharmacokinetics: modeling and simulation. In: Qiu Y, Chen Y, Zhang GGZ, Yu L, Mantri RV, editors. Developing solid oral dosage forms: pharmaceutical theory & practice, 2nd ed. London: Academic Press, 2017:399–413.
[18] Park M-N, Shin S-H, Byeon -J-J, Lee G-H, Yu B-Y, Shin YG. Prediction of pharmacokinetics and drug-drug interaction potential using physiologically based pharmacokinetic (PBPK) modeling approach: A case study of caffeine and ciprofloxacin. Korean J Physiol Pharmacol. 2017;21:107–15.
[19] Nakayama K, Ito S, Suzuki M, Takubo H, Yamazaki H, Nomura Y. Prediction of human pharmacokinetics of typical compounds by a physiologically based method using chimeric mice with humanized liver. Xenobiotica. 2019;49:404–14.
[20] Pires DEV, Blundell TL, Ascher DB. pkCSM: predicting small-molecule pharmacokinetic and toxicity properties using graph-based signatures. J Med Chem. 2015;58:4066–72.
[21] Barbara S. Rofecoxib (Vioxx) voluntarily withdrawn from market. Can Med Assoc J. 2004;171:1027–8.
[22] a) Ghosh J, Lawless MS, Waldman M, Gombar V, Fraczkiewicz R Modeling ADMET. In: Benfenati E, editor. In silico methods for predicting drug toxicity. New York: Hanuma Press, Springer Science+Business Media, 2016:63–83. (b) Mombelli E, Raitano G, Benfenati E. In silico prediction of chemically induced mutagenicity: how to use QSAR models and interpret their results. In: Benfenati E, editor. In silico methods for predicting drug toxicity. New York: Hanuma Press, Springer Science+Business Media, 2016:87–105; (c) Golbamaki A, Benfenati E. In silico methods for carcinogenicity assessment. In: Benfenati E, editor. In silico methods for predicting drug toxicity. New York: Hanuma Press, Springer Science+Business Media, 2016:107–19; (d) Marzo M, Roncaglioni A, Kulkarni S, Barton-Maclaren TS, Benfenati E. In silico model for developmental toxicity: how to use QSAR models and interpret their results. In: Benfenati E, editor. In silico methods for predicting drug toxicity. New York: Hanuma Press, Springer Science+Business Media, 2016:139–61; (e) Hewitt M, Przybylak K. In silico models for hepatotoxicity. In: Benfenati E, editor. In silico methods for predicting drug toxicity. New York: Hanuma Press, Springer Science+Business Media, 2016:201–36; (f) Roy K, Kar S. In silico models for ecotoxicity of pharmaceuticals. In: Benfenati E, editor. In silico

methods for predicting drug toxicity. New York: Hanuma Press, Springer Science+Business Media, 2016:237–304.

[23] Assessment and control of DNA reactive (mutagenic) impurities in pharmaceuticals to limit potential carcinogenic risk M7(R1): Current Step 4 Version 2017. http://www.ich.org/fileadmin/Public_Web_Site/ICH_Products/Guidelines/Multidisciplinary/M7/M7_R1_Addendum_Step_4_31Mar2017.pdf. Accessed: 15 Jan 2019.

[24] Patel M, Kranz M, Munoz-Muriedas J, Harvey JS, Giddings A, Swallow S, et al. A pharma-wide approach to address the genotoxicity prediction of primary aromatic amines. Comput Toxicol. 2018;7:27–35.

[25] Glück J, Buhrke T, Frenzel F, Braeuning A, Lampen A. In silico genotoxicity and carcinogenicity prediction for food-relevant secondary plant metabolites. Food Chem Toxicol. 2018;116:298–306.

[26] VEGA HUB. Virtual models for property Evaluation of chemicals within a Global Architecture. http://www.vega-qsar.eu/. Accessed: 15 Jan 2019.

[27] Lazar toxicity predictions. https://lazar.in-silico.ch/predict. Accessed: 15 Jan 2019; (a) Toxicity Estimation Software Tool (TEST). https://www.epa.gov/chemical-research/toxicity-estimation-software-tool-test. Accessed: 15 Jan 2019; (b) U.S. EPA (2016). User's Guide for T.E.S.T. (version 4.2) (Toxicity Estimation Software Tool): A Program to Estimate Toxicity from Molecular Structure.

[28] (a) Toxicity Estimation Software Tool (TEST). https://www.epa.gov/chemical-research/toxicity-estimation-software-tool-test. Accessed: 15 Jan 2019. (b) U.S. EPA (2016). User's Guide for T.E.S.T. (version 4.2) (Toxicity Estimation Software Tool): A Program to Estimate Toxicity from Molecular Structure.

[29] OECD QSAR ToolBox. http://www.oecd.org/chemicalsafety/risk-assessment/oecd-qsar-toolbox.htm. Accessed: 15 Jan 2019.

[30] Di Sotto A, Di Giacomo S, Abete L, Božović M, Parisi OA, Barile F, et al. Genotoxicity assessment of piperitenone oxide: an in vitro and in silico evaluation. Food Chem Toxicol. 2017;106(Pt A):506–13.

[31] a) Patlewicz G, Jeliazkova N, Safford RJ, Worth AP, Aleksiev B An evaluation of the implementation of the Cramer classification scheme in the Toxtree software. SAR QSAR Environ Res. 2008, 19, 495–524. b) Toxtree - Toxic hazard estimation by decision tree approach. http://toxtree.sourceforge.net/. Accessed: 17 Jan 2019.

[32] Bossa C, Benigni R, Tcheremenskaia O, Battistelli CL. (Q)SAR methods for predicting genotoxicity and carcinogenicity: scientific rationale and regulatory f frameworks. In: Nicolotti O, editor. Computational toxicology: methods and protocols, methods in molecular biology. Vol. 1800. New York: Springer Science+Business Media, 2018:447–73.

[33] ReFaC. http://www.refac.eu/aboutus.aspx. Accessed: 17 Jan 2019.

[34] European Food Safety Authority (EFSA). http://www.efsa.europa.eu/. Accessed: 17 Jan 2019.

[35] Verheyen GR, Deun KV, Van Miert S. Testing the mutagenicity potential of chemicals. J Genet Genome Res. 2017;4:029.

[36] Valerio LG, Jr., Cross KP. Characterization and validation of an in silico toxicology model to predict the mutagenic potential of drug impurities. Toxicol Appl Pharmacol. 2012;260:209–21.

[37] Plošnik A, Vračko M, Dolenc MS Mutagenic and carcinogenic structural alerts and their mechanisms of action. Arh Hig Rada Toksikol. 2016;67:169–82.

[38] (a) MultiCASE Inc. Beachwood, OH, (www.multicase.com). 2018. (b) Saiakhov R, Chakravarti, S, Klopman, G. Effectiveness of CASE Ultra expert system in evaluating adverse effects of drugs. Mol Inform. 2013;32,87–97.

[39] Beck DE, Abdelmalak M, Lv W, Reddy PVN, Tender GS, O'Neill E, et al. Discovery of potent indenoisoquinoline topoisomerase I poisons lacking the 3-nitro toxicophore. J Med Chem. 2015;58:3997–4015.

[40] Fujita Y, Honda H, Yamane M, Morita T, Matsuda T, Morita O. A decision tree-based integrated testing strategy for tailor-made carcinogenicity evaluation of test substances using genotoxicity test results and chemical spaces. Mutagen. 2019;34:101–9.
[41] The OECD QSAR Toolbox version 4.1. https://www.qsartoolbox.org. Accessed: 19 Jan 2019.
[42] Fan D, Yang H, Li F, Sun L, Di P, Li W, et al. In silico prediction of chemical genotoxicity using machine learning methods and structural alerts. Toxicol Res (Camb). 2017;7:211–20.
[43] PaDEL-Descriptor. A software to calculate molecular descriptors and fingerprints. http://www.yapcwsoft.com/dd/padeldescriptor/. Accessed: 19 Jan 2019.
[44] Dong J, Yao Z, Zhu M, Wang N, Lu B, Chen A, et al. ChemSAR: an online pipelining platform for molecular SAR modeling. J Cheminf. 2017;9:1–13.
[45] Orange 2.0 https://orange.biolab.si. Accessed: Jan 20, 2019.
[46] Ferrari T, Cattaneo D, Gini G, Bakhtyari NG, Manganaro A, Benfenati E. Automatic knowledge extraction from chemical structures: the case of mutagenicity prediction. SAR QSAR Environ Res. 2013;24:365–83.
[47] Morita T, Shigeta Y, Kawamura T, Fujita Y, Honda H, Honma M. In silico prediction of chromosome damage: comparison of three (Q)SAR models. Mutagen. 2019;34:91–100.
[48] Organisation for Economic Co-operation and Development (OECD). https://www.oecd-ilibrary.org/. Accessed: 20 Jan 2019.
[49] Lu J, Zhang P, Zou XW, Zhao XQ, Cheng KG, Zhao YL, et al. In silico prediction of chemical toxicity profile using local lazy learning. Comb Chem High Throughput Screen. 2017;20:346–53.
[50] Zhang H, Ren JX, Kang YL, Bo P, Liang JY, Ding L, et al. Development of novel in silico model for developmental toxicity assessment by using naïve Bayes classifier method. Reprod Toxicol. 2017;71:8–15.
[51] Hessel EVS, Staal YCM, Piersma AH. Design and validation of an ontology-driven animal-free testing strategy for developmental neurotoxicity testing. Toxicol Appl Pharmacol. 2018;354:136–52.
[52] López-Massaguer O, Pastor M, Sanz F, Carbonell P. Hepatotoxicity prediction by systems biology modeling of disturbed metabolic pathways using gene expression data. Methods Mol Biol. 2018;1800:505–18.
[53] Carbonell P, Lopez O, Amberg A, Pastor M, Sanz F. Hepatotoxicity prediction by systems biology modeling of disturbed metabolic pathways using gene expression data. ALTEX. 2017;34:219–34.
[54] Zhang L, Zhang H, Ai H, Hu H, Li S, Zhao J, et al. Applications of machine learning methods in drug toxicity prediction. Curr Top Med Chem. 2018;18:987–97.
[55] Banerjee P, Eckert AO, Schrey AK, Preissner R. ProTox-II: a webserver for the prediction of toxicity of chemicals. Nucleic Acids Res. 2018;46:W257–W263.
[56] ProTox-II – prediction of toxicity of chemicals. http://tox.charite.de/protox_II/. Accessed: 20 Jan 2019.
[57] Lu Y, Liu L, Lu D, Cai Y, Zheng M, Luo X, et al. Predicting hepatotoxicity of drug metabolites via an ensemble approach based on support vector machine. Comb Chem High Throughput Screen. 2017;20:839–49.
[58] Zhu XW, Li SJ. In silico prediction of drug-induced liver injury based on adverse drug reaction reports. Toxicol Sci. 2017;158:391–400.
[59] Ivanov S, Semin M, Lagunin A, Filimonov D, Poroikov V. In silico identification of proteins associated with drug-induced liver injury based on the prediction of drug-target interactions. Mol Inform. 2017;36:1600142.
[60] Toropova AP, Toropov AA. CORAL: binary classifications (active/inactive) for drug-induced liver injury. Toxicol Lett. 2017;268:51–7.

[61] CORAL-QSAR/QSPR. http://www.insilico.eu/coral/. Accessed: 21 Jan 2019.
[62] Kar S, Roy K, Leszczynski J. Impact of pharmaceuticals on the environment: risk assessment using QSAR modeling approach. In: Nicolotti O, editor. Methods in molecular biology - computational toxicology: methods and protocols. New York, NY: Humana Press, 2018:395–443.
[63] de Morais E, Silva L, Alves MF, Scotti L, Lopes WS, Scotti MT. Predictive ecotoxicity of MoA 1 of organic chemicals using in silico approaches. Ecotoxicol Environ Saf. 2018;153:151–9.
[64] Sangion A, Gramatica P. Hazard of pharmaceuticals for aquatic environment: prioritization by structural approaches and prediction of ecotoxicity. Environ Int. 2016;95:131–43.
[65] CASE Ultra Models. http://www.multicase.com/case-ultra-models#renaltox_bundle. Accessed: 21 Jan 2019.

Virendra R. Mishra, Chaitannya W. Ghanavatkar,
Vandana Kumari Shukla and Nagaiyan Sekar

7 Effect of substituent on photostability and lightfastness of azo dye and their photodegradation mechanism – Mechanistic study using density functional theory

Abstract: Density functional theory (DFT) derived global reactivity descriptor, i. e. electrophilicity index, was used to reveal that photostability of the azo dyes. Light fastness property of azo dyes increases with strong electron withdrawing substituents at the ortho position of azo group. Molecular electrostatic potential (MEP) plot were used to identify the reactivity of the molecule. Local reactivity descriptors particularly Fukui function were used to identify the reactive site for the attack of hydroxyl radical. Furthermore, it also explains that oxidative degradation of azo dye takes place through the cleavage of N-N bond. The photodegradation mechanism of azo dye in presence of hydroxyl radical was studied using total electronic energy of the optimized geometry for each step.

Keywords: density functional theory, azo dyes, substituents effect, electrophilicity index, MEP plots, Fukui function, photodegradation mechanism

This article has previously been published in the journal Physical Sciences Reviews. Please cite as: Mishra, V. R., Ghanavatkar, C. W., Shukla, V. K., Sekar, N. Effect of substituent on photostability and lightfastness of azo dye and their photodegradation mechanism – Mechanistic study using density functional theory *Physical Sciences Reviews* [Online] 2020, 5 DOI: 10.1515/psr-2019-0056.

https://doi.org/10.1515/9783110631623-007

Graphical Abstract:

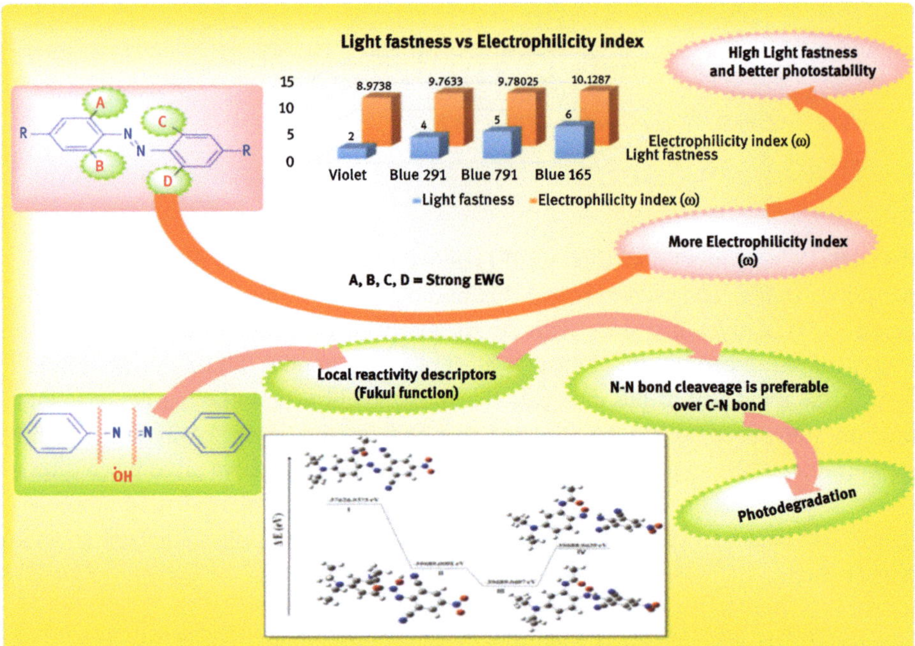

7.1 Introduction

Interaction of electromagnetic radiation on colored textiles causes several reactions and understanding this phenomenon is challenging for researchers [1]. The azo colorants are by far the most important class and are studied more than any other class in textile coloration [2, 3]. Colored polyester fibers have outlaid nylon fibers due to their good photostability and better lightfastness [4–6]. It is generally believed that dyeing of disperse dyes on polyester fiber will show high light fastness in comparison with the other substrates and that the polyester-disperse dye system is predominantly used where high light fastness is essential [7].

When colored fiber is exposed to electromagnetic radiation it brings change in hue or depth of color which depends on many factors such as the chemical structure of the colorant and substrate, the concentration of the colorant on the substrate, the energy distribution of the electromagnetic radiation, the amount of water, temperature, photostability of the dye and the presence of auxiliaries such as ultraviolet absorbing agent [8–10]. Enhancement of the lightfastness of azo dyes is an area of intense research [11, 12] and hence photodegradation mechanism of azo dye is widely studied [13–15]. Depending upon the characteristics of the polymer and the atmosphere, azo dye will either undergo photochemical oxidation or reduction reaction [16–21]. Photooxidation of an azo dye is facilitated by the active interaction with singlet oxygen producing nitrosobenzene, while in reductive environments photoreduction takes place which essentially produces a hydrazo compound and eventually, disproportionation giving substituted anilines. These two pathways have been dependably augmented in the literature [22, 23].

The influence of functional groups on the lightfastness of colorants is addressed by many researchers [24, 25]. However, there are confusing reports regarding the effect of functional groups attached to the azo component on the light fastness of azo dyes [26]. It has been proposed that electron deficient functional groups increase the photostability and/or lightfastness of an azo dye on polyester fabric [27]. Therefore, a good comprehension of the influence of substituent group on the photostability of azo dyes assumes more importance in deciding their proper end uses.

Peters et al. examined the photostability of azo dyes and correlated with some of the physical parameters namely pKa values of those dyes, and their behavior during mass spectrometry [28, 29]. The fading process of anthraquinone dyes with respect to their electronic properties can be carried out by using molecular orbital (MO) calculations [30]. To investigate the reaction of aminopyrazolinyl azo dyes with singlet molecular oxygen Morita and Hada used semi-empirical MO calculation [31]. Other studies indicate that electrophilic attack of singlet oxygen on an azo dye will enhance their photodegradation [32]. Density functional theory (DFT) helps to examine the site of photodegradation and effect of functional groups on photodegradation of azo dyes [33]. Semi-empirical (AM1) and *ab initio* methods were employed to determine the structure and electronic properties for a series of azobenzenes and azo thiophenes [34]. Several studies on all valence MO calculations, and particularly on azo dyes, are reported [35, 36] but it appears that an extensive investigation into the effect of substituents on light fastness of an azo dye needs to be carried out.

With a view of understanding the influence of functional groups on the photostability and light fastness of the azo dyes shown in Figure 7.1 DFT computations are performed and the results are reported in this paper. These dyes are chosen for investigation because of the ready availability of their lightfastness data and photodegradation mechanism in presence of hydroxyl radical on polyester and nylon fabrics [37]. The global reactivity descriptors are used to study the effect of different substituents on light fastness. Reactivity of the molecule and their photodegradation mechanism have been studied using molecular electrostatic potential (MEP) along with local reactivity descriptors. The result of this investigation will help to understand the effect of substituent on light fastness and photostability or photodegradation of an azo dye.

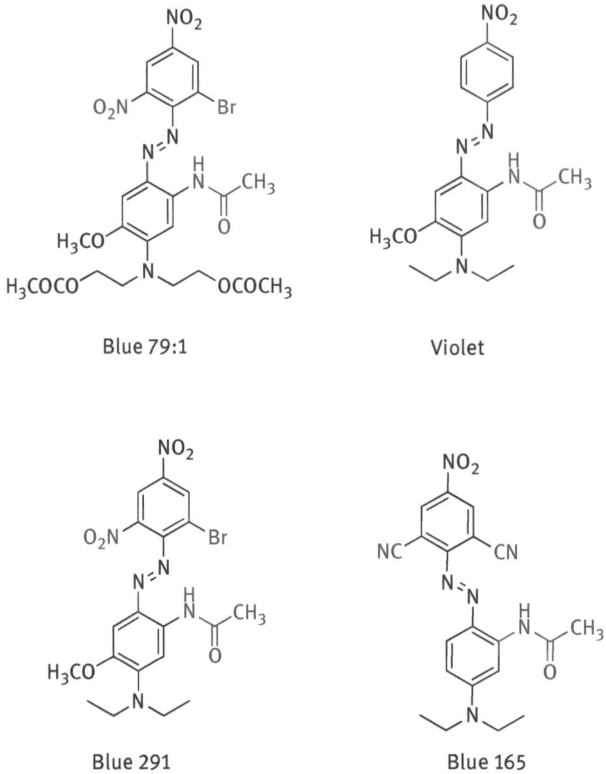

Figure 7.1: Monoazo disperse dyes.

7.2 Computational details

The computations were performed using the Gaussian 09 program package [38]. The ground state (S_0) geometry of the molecules under investigation was optimized as an isolated molecule in vacuum using the DFT method. The basis sets used for all atoms are the popular Pople's split-valence sets and the popular hybrid functional B3LYP, which combines Becke's three parameter exchange functional (B3) [39] with the non-local correlation functional by Lee, Yang and Parr (LYP) [40]. The B3LYP functional is used with the triple zeta basis set with both diffuse and polarization functions – 6-311++G(d, p) – for successive geometry optimization of azo dyes as reported in literature [41, 42]. The geometries optimized with B3LYP/6-311++G(d, p) method were used for further computations. To comprehend the behavior of azo dyes on polyester and nylon fabrics we have optimized the dyes in ester and amide solvents, respectively. Global reactivity descriptors are used to understand the influence of functional groups on light fastness and photostability of azo dyes. The mode of photodegradation mechanism was also studied using local reactivity descriptors.

7.3 Results and discussion

The geometries of all the dyes optimized at B3LYP/6-311++G(d, p) level of theory are given in Figure 7.2 in gas phase, ester solvent, i. e. methyl benzoate, butyl ethanoate, methyl methanoate, and amide solvents, i. e. formamide, *N*, *N*-dimethyl acetamide, *N*, *N*-dimethyl formamide, *N*-methylformamide mixture. The electronic energy of the highest occupied molecular orbital (E_{HOMO}) and the lowest unoccupied molecular orbital (E_{LUMO}) are used for further studies.

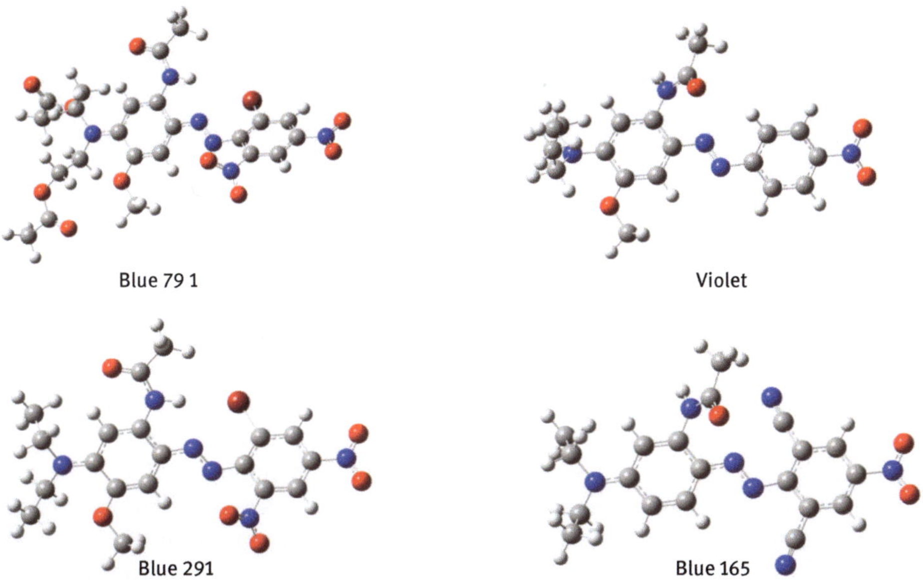

Figure 7.2: Optimized geometry of dyes in the gas phase at ground state using B3LYP/6-311++G(d, p).

7.4 Global reactivity descriptors

Global reactivity descriptors are used to study the effect of substituent on light fastness and photostability. On the basis of Koopman's theorem [43], global reactivity descriptors – electronegativity (χ), chemical potential (μ), chemical hardness (η), global softness (S), electrophilicity index (ω) – are calculated using the energies of Frontier Molecular Orbitals E_{HOMO}, E_{LUMO} given by eqs. (7.1–7.4) [44, 45].

$$\chi = \mu = -1/2\,(E_{HOMO} + E_{LUMO}) \tag{7.1}$$

$$\eta = 1/2(E_{LUMO} - E_{HOMO}) \tag{7.2}$$

$$S = 1/2\eta \tag{7.3}$$

$$\omega = \mu^2/2\eta \tag{7.4}$$

From Table 7.1, it is observed that the value of electrophilicity index (ω) for **Blue 165** is more than the value of ω for all the other dyes as shown in Figure 7.3 in both gas and solvents phase. The value of ω is more for **Blue 165**, as the azo group (-N = N-) is surrounded by stronger electron withdrawing group at the ortho position compared to all the other dyes. The molecule is stabilized more if the value of ω is more, and hence such molecules exhibit higher light fastness and better photostability [46, 47]. The trend in electrophilicity index (ω) is as follows: Blue 165 > Blue 79:1 > Blue 291 > Violet. A similar trend is observed when light fastness values are compared with electrophilicity index (ω) as mentioned in Table 7.2. Hence, Blue 165 exhibits higher light fastness value and more photostability than violet dye as shown in Figure 7.4.

7.5 Molecular electrostatic potential

MEP is an important method to elucidate the stable conformation, charge distribution, relative polarity and reactivity of a molecule [48, 49]. It is a valuable tool for qualitative elucidation of electrophilic and nucleophilic site of reaction. MEP simultaneously exhibits the molecular size, shape, positive, negative and neutral electrostatics potential region in terms of color grading. The different values of electrostatics potential at the surface are represented by different colors. The red, yellow and orange regions represent negative electrostatics potential and are related to electrophilic reactivity. Green and blue region represent positive electrostatics potential and are related to nucleophilic reactivity. The MEP plot for dyes are constructed from B3LYP/6-311++G(d, p) optimized geometry as shown in Figure 7.5.

7.6 Local reactivity descriptors

The azo dyes are mostly degraded through oxidative cleavage reaction in presence of hydroxyl radical [50]. A hydroxyl radical attack occurs either on the carbon connecting the azo group, leading to the breaking of the C-N bond or the nitrogen atom, followed by the breaking of the N-N bond as shown in Figure 7.6. The local reactivity descriptors, i. e. Fukui functions are used to study the oxidative cleavage of the azo dyes. Fukui functions (f_k^+, f_k^-, f_k^0) have been described earlier in literature [51, 52].

Using Mulliken population analyses of neutral, cation and anion state of the molecule, Fukui functions are calculated at the same level of theory B3LYP/6-311++G(d, p) in methyl methanoate and *N, N*-dimethyl formamide solvent using eqs. (7.5–7.7)

For nucleophilic attack

Table 7.1: HOMO and LUMO energies (eV), chemical potential (μ, eV), chemical hardness (η, eV), electrophilicity index (ω, eV), global softness (S, 1/eV) of all the dyes in gas and solvent phase at B3LYP/6-311++G(d, p).

Dyes	Opt file	HOMO Energy		LUMO Energy		μ	η	ω	S
		Hartee	eV	Hartee	eV	eV	eV	eV	1/eV
Blue 79:1	**Gas**	−0.2185	−5.9468	−0.1290	−3.5089	4.7278	1.2189	9.1689	0.4102
	Methyl benzoate	−0.2144	−5.8339	−0.1306	−3.5544	4.6941	1.1397	9.6664	0.4387
	Butyl ethanoate	−0.2147	−5.8420	−0.1305	−3.5503	4.6961	1.1459	9.6231	0.4363
	Methyl methanoate	−0.2142	−5.8284	−0.1308	−3.5579	4.6932	1.1353	9.7007	0.4404
	Formamide	−0.2134	−5.8075	−0.1311	−3.5663	4.6869	1.1206	9.8017	0.4462
	***N, N*-dimethyl acetamide**	−0.2135	−5.8107	−0.1310	−3.5647	4.6877	1.1230	9.7838	0.4452
	***N, N*-dimethyl formamide**	−0.2135	−5.8107	−0.1310	−3.5641	4.6874	1.1233	9.7803	0.4451
	***N*-methylformamide mixture**	−0.2134	−5.8069	−0.1311	−3.5671	4.6870	1.1199	9.8083	0.4465
Violet	**Gas**	−0.2122	−5.7737	−0.1170	−3.1840	4.4789	1.2949	7.7461	0.3861
	Methyl benzoate	−0.2090	−5.6864	−0.1223	−3.3282	4.5073	1.1791	8.6151	0.4241
	Butyl ethanoate	−0.2091	−5.6891	−0.1215	−3.3054	4.4972	1.1919	8.4847	0.4195
	Methyl methanoate	−0.2089	−5.6834	−0.1229	−3.3435	4.5134	1.1700	8.7059	0.4274
	Formamide	−0.2088	−5.6820	−0.1250	−3.4001	4.5410	1.1410	9.0366	0.4382
	***N, N*-dimethyl acetamide**	−0.2088	−5.6820	−0.1246	−3.3897	4.5359	1.1461	8.9754	0.4362
	***N, N*-dimethyl formamide**	−0.2088	−5.6820	−0.1246	−3.3895	4.5357	1.1463	8.9738	0.4362
	***N*-methylformamide mixture**	−0.2088	−5.6820	−0.1250	−3.4022	4.5421	1.1399	9.0496	0.4386
Blue 291	**Gas**	−0.2143	−5.8319	−0.1263	−3.4354	4.6337	1.1983	8.9593	0.4173
	Methyl benzoate	−0.2101	−5.7174	−0.1286	−3.4997	4.6085	1.1089	9.5767	0.4509
	Butyl ethanoate	−0.2103	−5.7234	−0.1283	−3.4912	4.6073	1.1161	9.5097	0.4480
	Methyl methanoate	−0.2100	−5.7133	−0.1289	−3.5062	4.6097	1.1036	9.6278	0.4531
	Formamide	−0.2096	−5.7043	−0.1297	−3.5285	4.6164	1.0879	9.7946	0.4596
	***N, N*-dimethyl acetamide**	−0.2097	−5.7054	−0.1295	−3.5244	4.6149	1.0905	9.7650	0.4585

(continued)

Table 7.1 (continued)

Dyes	Opt file	HOMO Energy		LUMO Energy		μ	η	ω	S
		Hartee	eV	Hartee	eV	eV	eV	eV	1/eV
	***N*, *N*-dimethyl formamide**	−0.2097	−5.7054	−0.1295	−3.5241	4.6148	1.0906	9.7632	0.4584
	***N*-methylformamide mixture**	−0.2096	−5.7041	−0.1297	−3.5296	4.6168	1.0872	9.8024	0.4599
Blue 165	**Gas**	−0.2273	−6.1846	−0.1331	−3.6221	4.9034	1.2812	9.3826	0.3902
	Methyl benzoate	−0.2219	−6.0374	−0.1348	−3.6676	4.8525	1.1849	9.9359	0.4220
	Butyl ethanoate	−0.2222	−6.0456	−0.1345	−3.6605	4.8530	1.1925	9.8746	0.4193
	Methyl methanoate	−0.2217	−6.0322	−0.1350	−3.6735	4.8529	1.1793	9.9846	0.4240
	Formamide	−0.2213	−6.0205	−0.1359	−3.6975	4.8590	1.1615	10.1630	0.4305
	***N*, *N*-dimethyl acetamide**	−0.2213	−6.0219	−0.1357	−3.6926	4.8572	1.1646	10.1287	0.4293
	***N*, *N*-dimethyl formamide**	−0.2213	−6.0219	−0.1357	−3.6926	4.8572	1.1646	10.1287	0.4293
	***N*-methylformamide mixture**	−0.2212	−6.0203	−0.1359	−3.6983	4.8593	1.1610	10.1690	0.4307

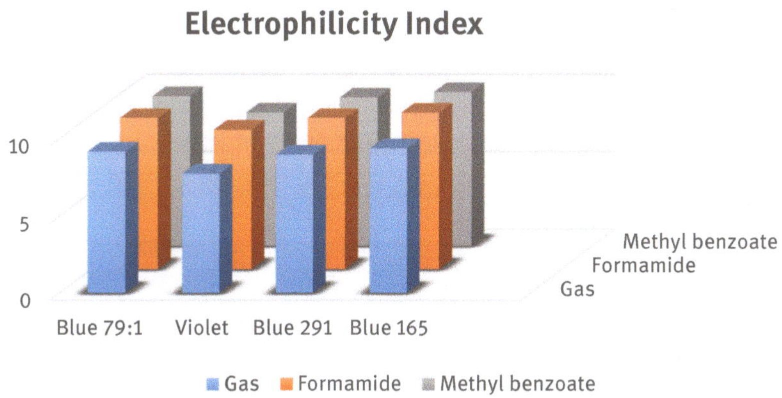

Figure 7.3: Electrophilicity index (ω) for all the dyes in gas and solvent phase (formamide and methyl benzoate) at B3LYP 6-311++G(d, p).

Table 7.2: Light fastness and Electrophilicity index (ω) for all dyes.

Dyes	Electrophilicity index (ω) eV		Light fastness
	Methyl methanoate	N, N-dimethyl formamide	
Blue 79:1	9.7007	9.7802	4–5
Violet	8.7059	8.9738	2
Blue 291	9.6278	9.7633	4
Blue 165	9.9846	10.1287	>6

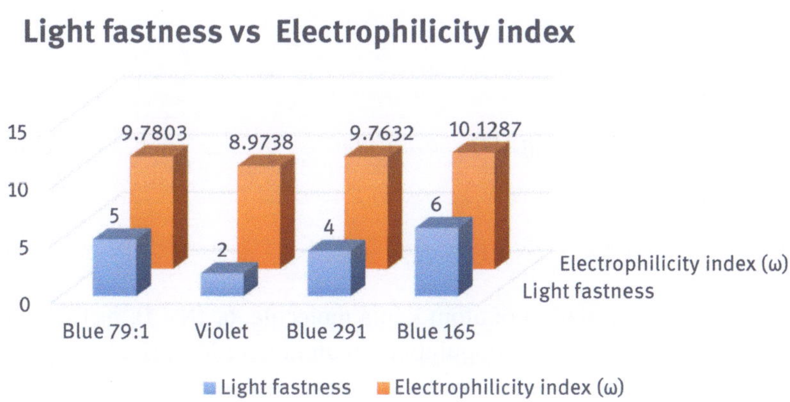

Figure 7.4: Comparison of light fastness and electrophilicity index (ω) (in N, N-dimethyl formamide solvent) for all the dyes.

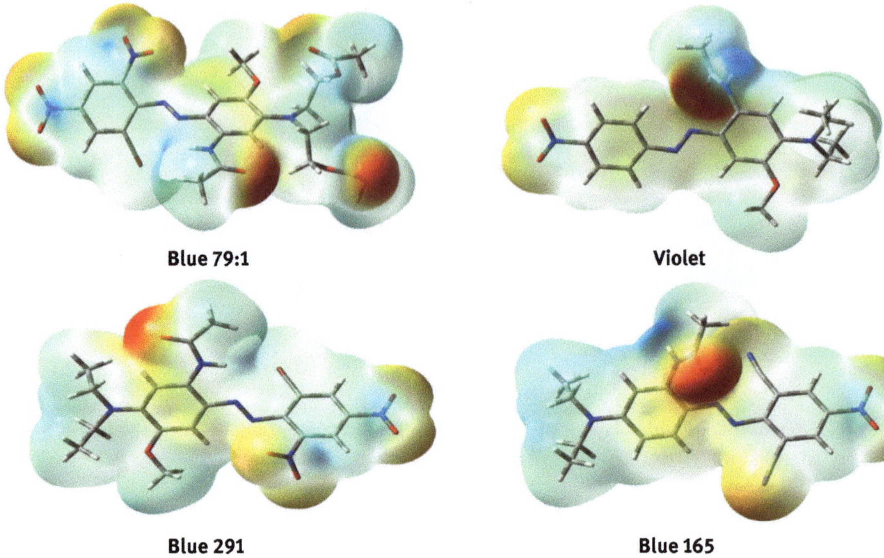

Figure 7.5: Molecular electrostatic potential of all dyes.

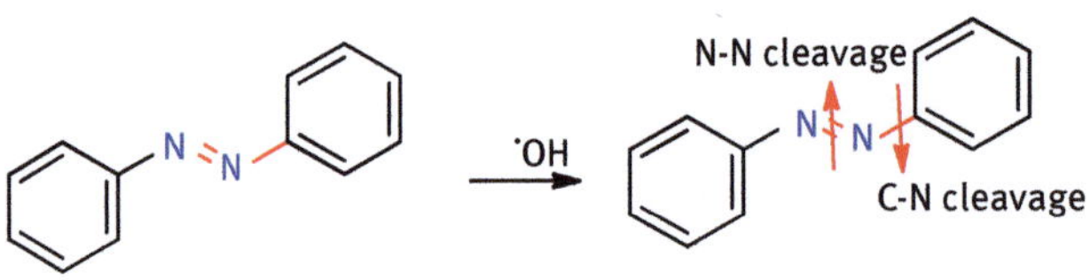

Figure 7.6: Oxidative cleavage of azo dyes.

$$f_k^{+} = [q_k(N+1) - q_k N] \tag{7.5}$$

For electrophilic attack

$$f_k^{-} = [q_k(N) - q_k(N-1)] \tag{7.6}$$

For radical attack

$$f_k^{0} = [q_k(N+1) - q_k(N-1)]/2 \tag{7.7}$$

where q_k is the electronic population of atom k in a molecule. q_k (N + 1), q_k (N) and q_k (N − 1) which refer to the electronic population on atom k for N + 1 (i. e. anionic), N (i. e. neutral), and N − 1 (i. e. cationic) electron systems, respectively.

Local softness and electrophilicity index are calculated using the following eqs. (7.8, 7.9)

$$s_k^{+} = Sf_k^{+},\ s_k^{-} = Sf_k^{-},\ s_k^{0} = Sf_k^{0} \quad (7.8)$$

$$\omega_k^{+} = \omega f_k^{+},\ \omega_k^{-} = \omega f_k^{-},\ \omega_k^{0} = \omega f_k^{0} \quad (7.9)$$

where +, –, 0 signs show nucleophilic, electrophilic and radical attack, respectively.

From Table 7.3, it was observed that hydroxyl radical will attack at nitrogen atom of the azo group; therefore, N-N bond cleavage was preferable over the C-N bond cleavage.

7.7 Photodegradation mechanism of Azo dyes

By calculating local reactivity descriptors, we recognized that the degradation of an azo dye will take place through radical attack on nitrogen atom of azo group (-N = N-). Hence, the degradation of the azo dye will take place by cleavage of N-N bond of azo group. From Fukui function (f_k^0) of **Blue 165** it is clear that hydroxyl radical will attack nitrogen atom N (11) of azo group (Figure 7.7). To understand the oxidative cleavage mechanism of N-N bond of azo group we have chosen the route mentioned in the literature (Figure 7.8) [53].

To recognize the mechanism of photodegradation of the azo dye in the presence of hydroxyl radical, we selected **Blue 165** which is optimized at B3LYP/6-311++G(d, p) in *N, N*-dimethyl formamide. Each step of the mechanism is optimized at the same level of theory. The photodegradation mechanism for Blue 165 in presence of hydroxyl radical is explained on the basis of the total electronic energy (ΔE) for each step of the mechanism. Total electronic energy of IV (–39,688.9629 eV) is less than that of I (–37,626.9374 eV), hence it is an exothermic reaction. Therefore, oxidative cleavage of N-N bond is the most promising route for photodegradation of **Blue 165** in presence of hydroxyl radical (Figure 7.9).

7.8 Conclusion

The purpose of this investigation was to examine the influence of functional groups on the photostability and light fastness of a few selected azo dyes, and thereby to elucidate the preferred reaction mechanism for oxidative cleavage of an azo dye in presence of hydroxyl radical by using DFT. Using global reactivity descriptor – electrophilicity index – we have understood the effect of ortho substituent on photostability and light fastness of the selected azo dyes. Dyes with a stronger electron withdrawing substituent at ortho position of an azo group exhibit more electrophilicity index, and hence show better light fastness and photostability. Reactivity of molecules was identified by using MEP plot. Local reactivity descriptors have shed light on to both the issues, i. e. site of reactivity and whether there is C-N or N-N

Table 7.3: Selected Fukui functions (f_k^0), local softness (s_k^0), local electrophilicity index (ω_k^0) for all dyes using Mulliken population analyses in Methyl methanoate and *N*, *N*-dimethyl formamide solvent at B3LYP/6-311++G(d, p).

Blue 79:1		f_k^0	s_k^0	ω_k^0	**Violet**		f_k^0	s_k^0	ω_k^0
Methyl methanoate	3 C	0.1020	0.0449	0.9899	**Methyl methanoate**	3 C	0.0416	0.0177	0.3623
	10 N	−0.2711	−0.1194	−2.6298		10 N	−0.0346	−0.0148	−0.3009
	11 N	0.1636	0.0720	1.5871		11 N	0.1426	0.0609	1.2414
	16 C	−0.2962	−0.1304	−2.8730		16 C	−0.1327	−0.0567	−1.1555
***N*, *N*-dimethyl formamide**	3 C	0.1036	0.0461	1.0137	***N*, *N*-dimethyl formamide**	3 C	0.0464	0.0202	0.4171
	10 N	−0.1398	−0.0622	−1.3673		10 N	−0.0437	−0.0191	−0.3924
	11 N	0.1591	0.0708	1.5563		11 N	0.1418	0.0618	1.2728
	16 C	−0.1454	−0.0647	−1.4217		16 C	−0.1320	−0.0576	−1.1845
Blue 291		f_k^0	s_k^0	ω_k^0	**Blue 165**		f_k^0	s_k^0	ω_k^0
Methyl methanoate	3 C	0.0982	0.0445	0.9458	**Methyl methanoate**	3 C	0.1257	0.0533	1.2558
	10 N	−0.1574	−0.0713	−1.5154		10 N	−0.1337	−0.0567	−1.3345
	11 N	0.1708	0.0774	1.6449		11 N	0.2010	0.0852	2.0072
	16 C	−0.1496	−0.0678	−1.4400		16 C	−0.1741	−0.0738	−1.7378
***N*, *N*-dimethyl formamide**	3 C	0.0995	0.0456	0.9723	***N*, *N*-dimethyl formamide**	3 C	0.1263	0.0542	1.2801
	10 N	−0.1609	−0.0738	−1.5707		10 N	−0.1360	−0.0584	−1.3771
	11 N	0.1654	0.0758	1.6155		11 N	0.1983	0.0851	2.0088
	16 C	−0.1459	−0.0669	−1.4243		16 C	−0.1695	−0.0727	−1.7164

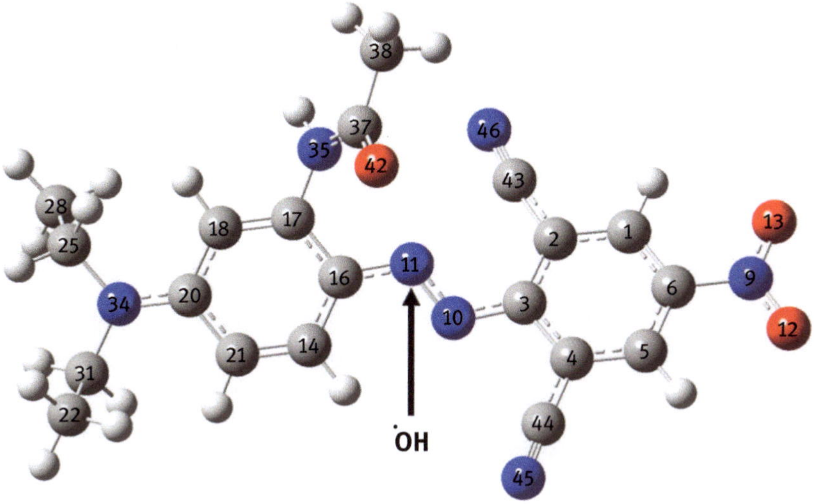

Figure 7.7: Optimized structure of **Blue 165** in *N*, *N*-dimethyl formamide solvent.

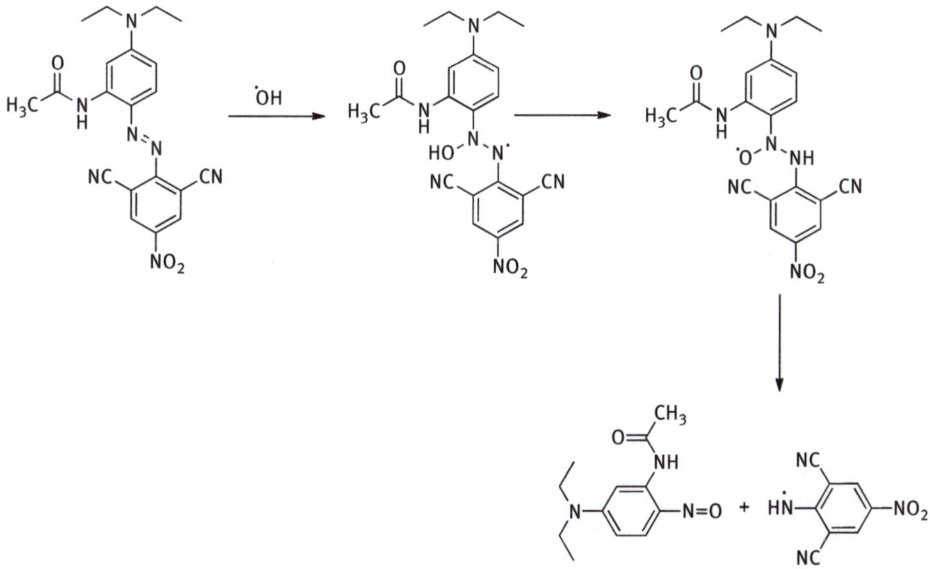

Figure 7.8: Oxidative cleavage mechanism of N-N bond of **Blue 165**.

bond cleavage in presence of hydroxyl radical. It is observed that N-N bond cleavage is more facile over C-N bond cleavage for all the dyes. By knowing the total electronic energy for each steps of mechanism we found that it is an exothermic reaction and it is more promising route for photodegradation of above azo dyes in presence of hydroxyl radical.

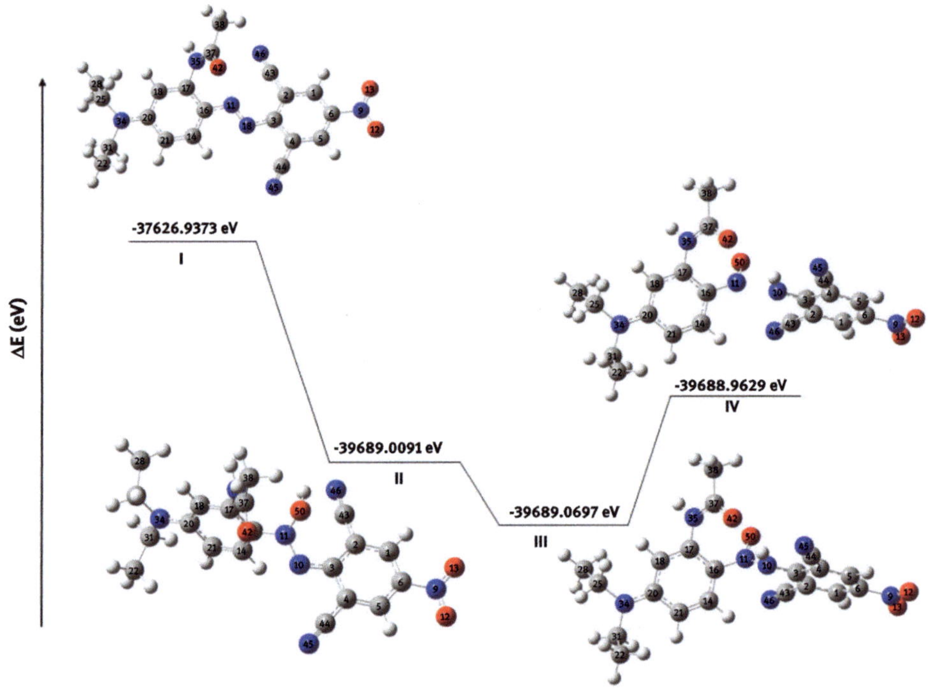

Figure 7.9: Energy profile diagram for degradation mechanism of **Blue 165** in presence of hydroxyl radical at the ground state using B3LYP/6-311++G(d, p).

Acknowledgements: Authors (VRM and CWG) acknowledge the office of the Principal Scientific Adviser to Government of India for financial support by way of Junior Research Fellowship.

References

[1] Ascension R. Modeling the effects of color on the UV protection provided by cotton woven fabrics dyed with azo dyestuffs. Ind Eng Chem Res. 2009;48:9817–22.

[2] Hunger K. Industrial dyes: chemistry, properties, applications. 2003.

[3] Brüschweiler BJ. Azo dyes in clothing textiles can be cleaved into a series of mutagenic aromatic amines which are not regulated yet. Regul Toxicol Pharmacol. 2017;88:214–26.

[4] Shah TB, Shiny RS, Dixit RB, Dixit BC. Synthesis and dyeing properties of new disazo disperse dyes for polyester and nylon fabrics. J Saudi Chem Soc. 2014;18:985–92.

[5] Otutu JO, Okoro D. Preparation of dis-azo dyes derived from p-aminophenol and their fastness properties for synthetic polymer-fibres. J Appl Sci. 2008;8:334–9.

[6] Allen NS, Harwood B, McKellar JF. Lightfastness and spectroscopic properties of amino-chloroanthraquinones. J Photochem. 1978;9:559–64.

[7] Ho YW. Synthesis of some new azo pyrazolo [1, 5- a] pyrimidine-thieno [2, 3- b] pyridine derivatives and their application as disperse dyes. Dye Pigm. 2005;64:223–30.

[8] Giles CH, Mckay RB. The lightfastness of dyes: a review. Text Res J. 1963;33:528–77.
[9] Baxter G, Glies C. The influence of the physical state of dyes upon their light fastness. Color Technol. 1954;71:218–35.
[10] Millington KR, Ponds W Colorfastness. Engineering of High-Performance Textiles; 2018.
[11] Freeman HS, Posey JC. An approach to the design of lightfast disperse dyes- analogs of disperse yellow 42. Dye Pigm. 1992;20:171–95.
[12] Hallas G, Choi JH. Synthesis and properties of novel aziridinyl azo dyes from 2-aminothiophenes-Part 2: application of some disperse dyes to polyester fibers. Dye Pigm. 1999;40:119–29.
[13] Saleh SM. ZnO nanospheres based simple hydrothermal route for photocatalytic degradation of azo dye. Spectrochim Acta Part A Mol Biomol Spectrosc. 2019;211:141–7.
[14] Abbasi A, Najafi M, Janczak J, Hecke KV. Mo (VI) and W (VI) complexes as heterogeneous catalysts for degradation of azo dyes. J Environ Chem Eng. 2019;7:1.
[15] Hassani KEl, Kalnina D. Enhanced degradation of an azo dye by catalytic ozonation over Ni-containing layered double hydroxide nanocatalyst. Sep Purif Technol. 2019;210:764–74.
[16] Feng W, Nansheng D, Helin H. Degradation mechanism of azo dye C. I. reactive red 2 by iron powder reduction and photooxidation in aqueous solutions. Chemosphere. 2000;41:1233–8.
[17] He J, Ma W, Zhao J. Photooxidation of azo dye in aqueous dispersions of H2O2/α-FeOOH. Appl Catal B Environ. 2002;39:211–20.
[18] Bauer C, Jacques P, Kalt A. Photooxidation of an azo dye induced by visible light incident on the surface of TiO_2. J Photochem Photobiol A Chem. 2001;140:87–92.
[19] Allen NS. Photofading mechanisms of dyes in solution and polymer media. Rev Prog Color Relat Top. 1987;17:61–71.
[20] Vig A, Sallay P, Sirbiladze K, Nagy HJ, Aranyosi P, Ruszna I. The light stability of azo dyes and dyeings V. The impact of the atmosphere on the light stability of dyeings with heterobifunctional reactive azo dyes. Dye Pigm. 2007;72:16–22.
[21] Dong Y, Dong W, Liu C, Chen Y, Hua J. Photocatalytic decoloration of water-soluble azo dyes by reduction based on bisulfite-mediated borohydride. Catal Today. 2007;126:456–62.
[22] Freeman HS, Hsu WN. Photolytic behavior of some popular disperse dyes on polyester and nylon substrates. Text Res J. 1987;57:223–34.
[23] Lishan DG, Harris PW, Brahim K, Jackson RL. Photodegradation of acid dyes in nylon films. J Soc Dye Colour. 1988;104:33–7.
[24] Kienle RH, Stearns El, Meulen PA. A correlation of structure with fastness to light of monoazo dyes. J Phys Chem. 1946;50:363–73.
[25] Chipalkatti HR, Desai NF, Giles CH, Macaulay N. The influence of the substrate upon the light fading of azo dyes. J Soc Dye Colour. 1954;70:487–501.
[26] Rageh NM, Abdallah EM. Spectrophotometric investigation of the role of organic solvents in the ionization of azo dyes derived from thienopyridine. J Chem Eng Data. 2003;48:1495–9.
[27] Giles CH, Bernat J. The nature of azo dye light fading in polyester and the physical state of the dye. Text Res J. 1976;46:673–5.
[28] Sunthankar SU, Thanumoorthy V. Disperse dyes: dyeing properties and fastness of azo dyes. Text Res J. 1977;47:126–8.
[29] Peters AT, Bridgeman I. Photochemical degradation of aminoazobenzene disperse dyes in ethanolic solution part ii: effect of pH and of dye structure. Text Res J. 1973;44:645–9.
[30] Ray A, Deheri GM. CNDO/II study on the fading of anthraquinone dyes. Dye Pigm. 1995;27:327–31.
[31] Morita Z, Hada S. A semiepirical molecular orbital study on the reaction of an aminopyrazolinyl azo dye with singlet molecular oxygen. Dye Pigm. 1999;41:1–10.

[32] Guo L, Meng F, Gong X, Xiao H, Chen K, Tian H. Synthesis and spectral properties of soluble trimethylsilyl substituted metal-phthalocyanines. Dye Pigm. 2001;49:83–91.
[33] Özen AS, Aviyente V, Klein RA. Modeling the oxidative degradation of azo dyes: a density functional theory study. J Phys Chem A. 2003;107:4898–907.
[34] Özen AS, Aviyente V, De Proft F, Geerlings P. Modeling the substituent effect on the oxidative degradation of azo dyes. J Phys Chem A. 2004;108:5990–6000.
[35] Stoilova A, Georgiev A, Nedelchev L, Nazarova D, Dimov D. Structure-property relationship and photoinduced birefringence of the azo and azo-azomethine dyes thin films in PMMA matrix. Opt Mater. 2019;87:16–23.
[36] Madkour LH, Guo L, Kaya C. Quantum chemical calculations, molecular dynamic (MD) simulations and experimental studies of using some azo dyes as corrosion inhibitors for iron. Part 2: bis e azo dye derivatives. J Mol Struct. 2018;1163:397–417.
[37] Himeno K, Okada Y, Morita Z. Photofading of monoazo disperse dyes on polyester and polyamide substrates. Dye Pigm. 2000;45:109–23.
[38] Frisch MJ, Trucks GW, Schlegel HB, et al. Gaussian 09, Revision C.01, Inc. Wallingford, CT, 2009
[39] Becke AD. A new mixing of Hartree–Fock and local density functional theories. J Chem Phys. 1993;98:1372–7.
[40] Lee C, Yang W, Parr RG. Development of the Colic-Salvetti correlation-energy into a functional of the electron density formula Chengteh. Phys Rev B. 1988;37:785–9.
[41] Cinar M, Coruh A, Karabacak M. A comparative study of selected disperse azo dye derivatives based on spectroscopic (FT-IR, NMR and UV – vis) and nonlinear optical behaviors. Spectrochim Acta Part A Mol Biomol Spectrosc. 2014;122:682–9.
[42] Tathe AB, Sekar N. Red emitting coumarin—azo dyes: synthesis, characterization, linear and non-linear optical properties-experimental and computational approach. J Fluoresc. 2016;26:1279–93.
[43] De Proft F, Geerlings P. Conceptual and computational DFT in the study of aromaticity. Chem Rev. 2001;101:1451–64.
[44] Chattaraj PK, Chakraborty A, Giri S. Net electrophilicity. J Phys Chem A. 2009;113:10068–74.
[45] Parr RG, Szentpály LV, Liu S. Electrophilicity index. J Am Chem Soc. 1999;121:1922–4.
[46] Bhide R, Jadhav AG, Sekar N. Light fast monoazo dyes with an inbuilt photostabilizing unit: synthesis and computational studies. Fibers Polym. 2016;17:349–57.
[47] Mishra VR, Sekar N. Photostability of coumarin laser dyes -a mechanistic study using global and local reactivity descriptors. J Fluoresc. 2017;27:1101–8.
[48] Serdaro G, Uluda N. A novel synthesis of octahydropyrido [3, 2- c] carbazole framework of aspidospermidine alkaloids and a combined computational, FT-IR, NMR, NBO, NLO, FMO, MEP study of the cis –4a-Ethyl-1- carbazole. J Mol Struct. 2018;1161:152–68.
[49] Soliman SM, Hagar M, Ibid F, Sayed E, El AH. Experimental and theoretical spectroscopic studies, HOMO – LUMO, NBO analyses and thione – thiol tautomerism of a new hybrid of 1, 3, 4-oxadiazole-thione with quinazolin-4-one. Spectrochim Acta Part A Mol Biomol Spectrosc. 2015;145:270–9.
[50] Alimet SO, Klein RA. Modeling the oxidative degradation of azo dyes: a density functional theory study. J Phys Chem A. 2003;107:4898–907.
[51] Rawat P, Singh RN. Experimental and theoretical study of 4-formyl pyrrole derived aroylhydrazones. J Mol Struct. 2015;1084:326–39.
[52] Singh RN, Kumar A, Tiwari RK, Rawat P, Manohar R. Synthesis, molecular structure, and spectral analyses of ethyl-4-[(2,4-dinitrophenyl)-hydrazonomethyl]-3,5-dimethyl-1H-pyrrole-2-carboxylate. Struct Chem. 2012;24:713–24.
[53] Alimet SO, De Proft F, Geerlings P. Modeling the substituent effect on the oxidative degradation of azo dyes. J Phys Chem. 2004;108:5990–6000.

R. Kavipriya, Helen P. Kavitha, B. Karthikeyan, Jasmine P. Vennila, Lydia Rhyman and Ponnadurai Ramasami

8 2,4-Dimorpholino-4-yl-6-(4-nitrophenoxy)-[1,3,5]-triazine: Structural and spectroscopic study using experimental and DFT method

Abstract: 2,4-Dimorpholino-4-yl-6-(4-nitrophenoxy)-[1,3,5]-triazine (DMNT) was synthesized and the molecular structure and vibrational frequencies were studied by density functional theory (DFT) method. The functional used was Becke's three parameter exchange functional combined with the Lee-Yang-Parr correlation (B3LYP) and the standard basis set was 6-31G(d) for all atoms. The Fourier Transform-Infra Red (FT-IR) and FT-Raman spectra of DMNT were recorded and complete assignments of the observed vibrational frequencies are done. The assignments were confirmed by isotopic labelling. The structural parameters, harmonic vibrational frequencies, IR intensities and Raman intensities of DMNT in the ground-state were also computed. Non-linear optical behaviour of DMNT was analysed by examining the properties like electric dipole moment, polarizability and hyperpolarizability. Molecular properties such as ionization potential, electro-negativity, chemical potential and chemical hardness were obtained from molecular orbital analysis. Hyper conjugative interaction and charge delocalization taking place in DMNT was confirmed by Natural bond analysis studies. UV-Vis spectrum of DMNT was also recorded to understand the electronic properties.

Keywords: 2,4-Dimorpholino-4-yl-6-(4-nitrophenoxy)-[1,3,5]-triazine, DFT, HOMO, LUMO, NBO, vibrational spectra

This article has previously been published in the journal Physical Sciences Reviews. Please cite as: Kavipriya, R., Kavitha, H. P., Karthikeyan, B., Vennila, J. P., Rhyman, L., Ramasami, P. 2,4-Dimorpholino-4-yl-6-(4-nitrophenoxy)-[1,3,5]-triazine: Structural and spectroscopic study using experimental and DFT method *Physical Sciences Reviews* [Online] 2019, 4 DOI: 10.1515/psr-2019-0003.

https://doi.org/10.1515/9783110631623-008

Graphical Abstract:

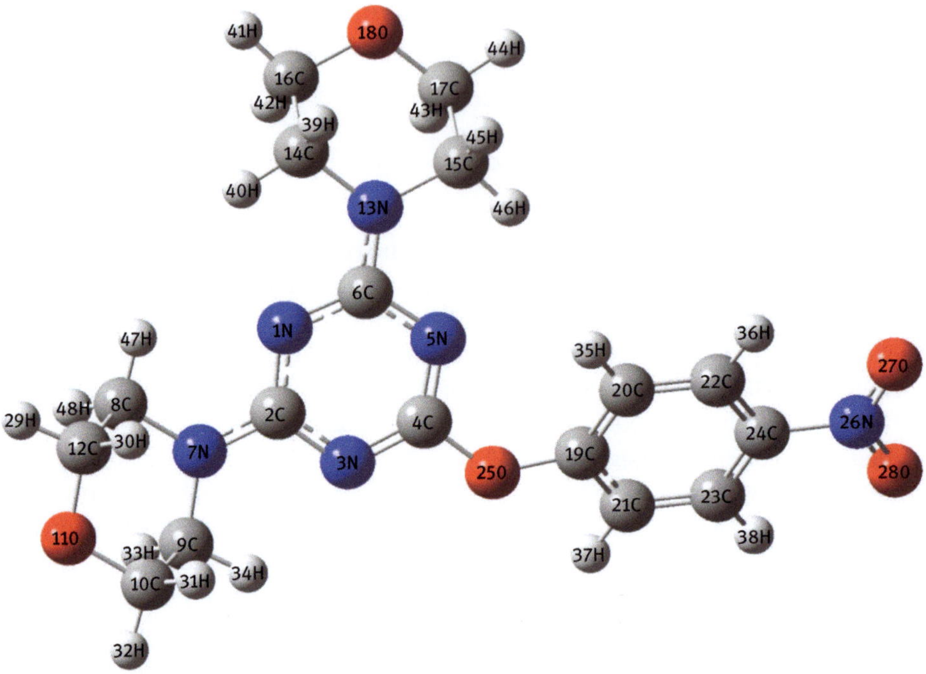

8.1 Introduction

Triazines are the potential derivatives of heterocyclic compounds possessing nitrogen. Compounds containing 1,3,5-triazines are present as functional site in many bioactive compounds which are obtained by natural [1–3] and synthetic methods [4–6]. These heterocyclic compounds readily go through ring transformations and substitution reactions [7, 8] and hence they lead to the synthesis of other heteroaromatic compounds. Triazines are subjected to various biotic and abiotic degradation processes and quantification of these metabolic products provides an additional analytical tool to check contamination in water [9]. 1,3,5-Triazines possess a broad variety of biological and pharmacological applications like anti-cancer [10], anti-ulcer [11] and anti-inflammatory [12] effects. They also play an important role in agriculture in regulating the growth of the crops [13–15]. The FT-IR and FT-Raman spectra of triazine and its halogenated derivatives have been reported both experimentally and theoretically [16]. Marchewka reported that the derivatives of melamine possess non-linear optical (NLO) properties [17].

In view of the importance of triazines, using experimental and density functional theory method, we investigated 2,4-dimorpholino-4-yl-6-(4-nitrophenoxy)-[1,3,5]-triazine (DMNT). It is confirmed from the literature, that no study about DMNT has been made so far and this is the first detailed report of DMNT. The objectives of the study are (i) to synthesize DMNT (ii) to obtain the Fourier Transform-Infra Red (FT-IR) and FT-Raman spectra of DMNT (iii) to obtain the optimized geometrical parameters (iv) to predict bioactive sites (v) to study properties such as dipole moment, polarizability and hyperpolarizability and to explore their NLO behaviour (vi) to find the molecular orbitals – highest occupied molecular orbital (HOMO) and lowest unoccupied molecular orbital (LUMO) (vii) to perform natural bond orbital (NBO) analysis (viii) to study the variation of thermodynamic parameters with temperature and (ix) to understand the electronic property of DMNT.

8.2 Synthesis and experimental methods

A slurry of cyanuric chloride is obtained when a solution of cyanuric chloride (0.01 mol) in 8 mL acetone was added with stirring to a cold solution of sodium bicarbonate (0.01 mol) in 10 mL distilled water. A solution of nitrophenol (0.01 mol) in 10 mL acetone was added to it and a saturated solution of sodium bicarbonate was added to neutralize the reaction mixture and it was stirred for 2 h at 0–5 °C. Then a mixture of sodium hydroxide (0.02 mol) and morpholine (0.01 mol) in 8 mL distilled water was added slowly and the reaction was continued for 4 h. The resultant product obtained was filtered and re-crystallized from ethanol. The FT-IR and FT-Raman spectrum of DMNT was recorded in Perkin-Elmer 180 Spectrometer.

8.3 Quantum chemical computations

Gaussian 03 W package [18] was used for all computations of the DMNT compound in the gas phase with B3LYP/6–31G(d) basis set. By considering DMNT in the C_s point group symmetry, the optimized structural parameters were calculated for the optimized geometry. Gaussview Program [19] was used for visual animation of the vibrational modes and their assignment. The possibility of charge transfer in the compound is obtained from the molecular orbital analysis (HOMO and LUMO). Energy gap between the molecular orbitals gives details about the chemical hardness (η) and chemical potential (μ). Molecular parameters such as dipole moment and hyperpolarizability reveals the NLO activity of the compound. Intermolecular charge transfer (ICT) in DMNT is obtained from NBO analysis.

8.4 Results and discussion

8.4.1 Molecular geometry

Table 8.1 represents the optimized structural parameters of the DMNT and the Figure 8.1 depicts the atom numbering scheme. The optimized geometry of DMNT is compared with other similar triazine derivaties with reported X-ray structures as the crystallographic data of the DMNT is not analysed so far. The C-N ring bond distances in the optimized geometry of DMNT are 1.345, 1.342, 1.357, 1.360, 1.322 and 1.354 Å and the side chains C-N have the bond length 1.360 Å. They match with the respective bond distances obtained from the XRD data [20]. It is reported that the

Table 8.1: Optimized geometrical structural parameters of DMNT using B3LYP/6–31 G(d) method.

Parameters	Method	Exptl.
Bond length (Å)		
N1-C2	1.345	1.345
N1-C6	1.342	1.340
C2-N3	1.354	1.354
C2-N7	1.360	1.356
N3-C4	1.322	1.319
C4-N5	1.323	1.327
C4-O25	1.365	1.358
N5-C6	1.357	1.357
C6-N13	1.360	
N7-C8	1.460	1.457
N7-C9	1.462	
C8-C12	1.530	1.500
C8-H47	1.089	0.970
N26-O27	1.232	1.244
C9-H33	1.100	
C14-C16	1.530	1.534
Bond angle (°)		
C2-N1-C6	115.2	115.96
N1-C2-N3	124.9	125.6
N1-C2-N7	118.0	118.0
N3-C2-N7	116.9	117.5
C2-N3-C4	113.3	
N3-C4-O25	112.7	111.74
N5-C4-O25	118.6	118.40
C4-N5-C6	113.3	
N7-C8-C12	109.4	108.5
N7-C9-H33	109.1	109.7
C12-C8-H47	111.2	111.5

Note: XRD values are taken from Reference [20].

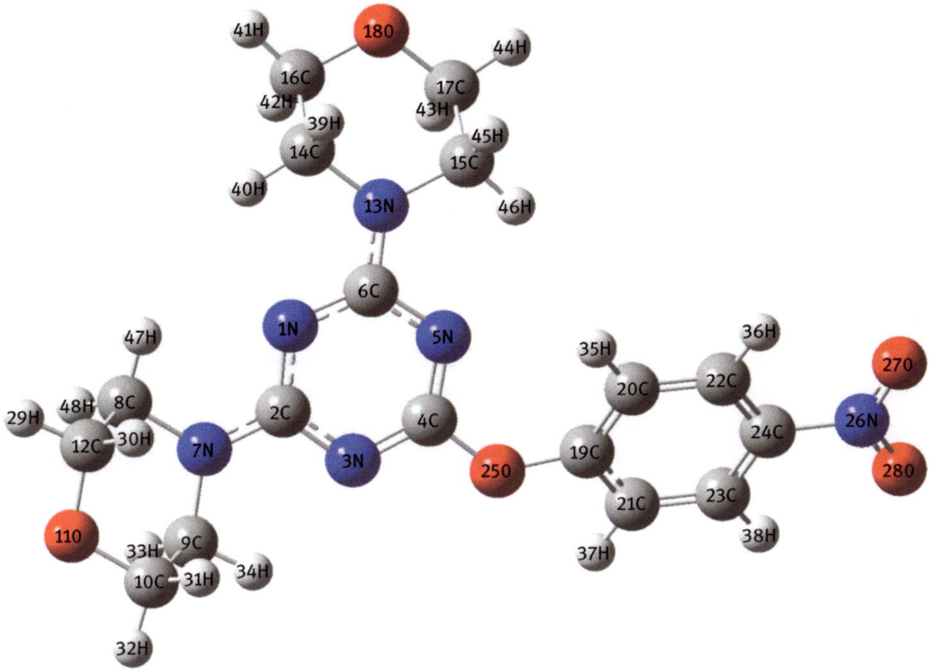

Figure 8.1: Optimized molecular structure of DMNT.

changes in the substitutions on the carbon atom of the aromatic ring alters the length of the C-H bond [21]. In the present compound DMNT, the electron density at the ring carbon atoms is reduced due to the presence of electron withdrawing NO_2 group. In substituted benzenes, the ring carbon atom exhibits a decrease in the corresponding bond length as it has larger attraction on the valence electron cloud of the hydrogen atom [21].

The calculated bond length for C-C ranges from 1.374 to 1.387 Å and C-H ranges from 1.064 to 1.068 Å. The resonating structures of the nitro group confirms the delocalization $O{=}N^{+}{-}O^{-}$ and $O{-}N^{+}{=}O$ with the bond length of the N-O bonds (1.443 Å) . The order C24-C23-C21 < C19-C20-C22 < C20-C22-C24 < C23-C19 < C21-C19-C20 < C22-C24-C23 in the benzene ring predicts its asymmetry. The bond angle C22-C24-C23 (121.2) is 0.5° greater than the bond angle C20-C19-C21 (120.7°). The substitution of the electron withdrawing nitro group on C24 increases the bond angle of carbon C24.

8.4.2 Normal coordinate analysis

In DMNT compound, there are 138 fundamental modes of vibration in agreement with C_s point group of symmetry which includes 98 in-plane and 40 out-of-plane vibrations. Detailed descriptions of the vibrational modes were given with the help of normal coordinates. The full sets of 124 internal coordinates (with 26 redundancies) were identified and given in Table SI1. Among this, a non redundant set of local symmetry coordinates (Table SI2) were noted by means of suitable linear combinations of internal coordinates as given by Fogarasi and Pulay [22, 23].

8.4.3 Vibrational assignments

The important vibrational modes of assignment with the experimentally observed and theoretically calculated frequencies are given in Table 8.2. The calculated Raman activities (S_i) were converted to relative Raman intensities (I_i) using the basic theory of Raman scattering [24–27]. The potential energy distributions (PED) for DMNT were calculated by using VEDA 4 program [28] and the fundamental vibrational modes were characterized by their PED's. The experimental FT-IR and FT-Raman spectra are represented in Figure 8.2 and Figure 8.3.

8.4.3.1 Nitro group vibrations

There are two bands in the compounds containing nitro group namely asymmetric stretching and symmetric stretching which occur at 1540–1614 cm^{-1} and 1260–1390 cm^{-1}, respectively [29]. The exact position of the bands depends on the unsaturation near the nitro group and upon the type of substituent near to it. The in-plane vibrations of nitro group in aromatic compounds occur in the region 590–500 cm^{-1} [30]. The other vibrations of NO_2 (rocking, wagging, scissoring and twisting) occur in the low-frequency region [31]. For DMNT, the band observed at 1583 cm^{-1} in FT-IR corresponds to NO_2 asymmetric vibrations and it agrees with the scaled value at 1566 cm^{-1}. NO_2 symmetric vibrations are noted at 1300 cm^{-1} in FT-IR and it is in good agreement with the scaled frequency at 1300 cm^{-1}. Owing to the strong resonance between the phenyl ring and the NO_2 group, the symmetrical NO_2 vibration appears at lower frequencies with increased intensity. NO_2 scissoring is observed at 752 cm^{-1} in FT-IR spectrum. The band at 543 cm^{-1} in FT-IR and 544 cm^{-1} in FT-Raman is assigned to NO_2 wagging and the FT-IR band at 462 cm^{-1} is assigned to NO_2 rocking [24]. NO_2 torsional mode appear below 100 cm^{-1} and this is possible only in FT-Raman spectrum. In the compound DMNT, the peak obtained at 71 cm^{-1} in FT-Raman corresponds to NO_2 torsion.

8.4.3.2 C-N vibrations

In a compound containing nitro group, the C-N stretching and C-N bending vibrations appear at nearly 870 and 610 cm^{-1}, respectively [32]. The C-N stretching of the nitro group of DMNT appears at 856 cm^{-1} in FT-IR and 859 cm^{-1} in FT-Raman and the C-N bending vibrations occur at 505 cm^{-1} in FT-IR. These assignments agree

Table 8.2: The calculated and observed FT-IR and FT-Raman spectrum of DMNT molecule by B3LYP/6-31G(d)method.

Mode nos	Experimental (cm^{-1})		Calculated (cm^{-1})	Scaled (cm^{-1})	μ	F	I_{IR}	I_{Raman}	Vibrational assignments
	FT-IR	FT-Raman							
7		71	69	68	5.1276	0.0145	0.0817	0.1810	γ CCNO (64)
9		95	101	100	4.0245	0.0243	0.2401	0.0317	γ CNCN(58), γ NNNC(13)
27	462		459	455	4.2079	0.5225	2.6803	0.0857	ρCCNO (11)
30	505		512	508	3.1115	0.4809	3.1501	0.3437	β CNO (62)
32	543	544	544	540	6.8286	1.1927	7.3733	0.3246	ωNO_2 (24)
36		640	636	631	5.9423	1.4199	10.5510	0.0262	β NCN (16)
37	675		679	674	2.1247	0.5785	22.0596	0.0163	γ CCCC (46)
41	752		748	742	5.4879	1.8101	42.6535	0.3250	δ NO_2 (31)+ γ C-H(12)
43	800	830	843	836	8.9203	3.7389	3.0275	0.4239	ρCH_2 (51)
45	856	859	867	860	4.3934	1.9462	20.9687	0.3134	$_v$C-N (43)
55		972	982	974	2.7736	1.5765	1.5917	0.0385	γ C-H (11)
57	1018		1002	994	1.2492	0.7394	0.9121	0.1302	$_v$ C-O-C (82)
62	1112	1109	1120	1111	2.4644	1.8244	86.1490	0.0472	tCH_2 (60)
65	1166	1170	1186	1177	1.7002	1.4101	6.7593	0.0103	β CCC (40)
76	1265		1263	1253	2.8374	2.6709	11.6556	0.1038	ωCH_2 (54)
79	1300		1308	1298	1.6174	1.6328	335.0848	3.6660	$_v$N-O (14)
92	1452		1463	1451	1.6874	2.1300	46.9528	0.4883	tCH_2 (75)
93		1463	1471	1459	1.4227	1.8155	12.5240	0.1294	βC-H (13)
94	1490		1506	1494	1.2304	1.6442	6.8558	0.0263	βC-H (16)
96		1503	1515	1503	1.2084	1.6356	21.0684	0.0217	$_\upsilon$ C-N (58)
99		1525	1533	1521	1.4433	1.9996	183.0209	0.0880	ωCH_2 (62)
103	1583		1579	1566	2.9020	4.2639	70.7848	0.3470	$_{\upsilon a}$ NO_2 (61)
104		1595	1615	1602	5.9132	9.0910	1090.2876	0.3331	$_\upsilon$ C-C (68)
106			1649	1636	1.0916	1.7492	3.2574	14.2253	$_v$ C-N (29)

(continued)

Table 8.2 (continued)

Mode nos	Experimental (cm^{-1})		Calculated (cm^{-1})	Scaled (cm^{-1})	μ	F	I_{IR}	I_{Raman}	Vibrational assignments
	FT-IR	FT-Raman							
109			1659	1646	1.0874	1.7637	2.8376	30.9853	$_{υ}$ C-C (28)
110			1660	1647	1.0916	1.7714	9.2763	11.7163	$_{υ}$ C-C (34)
112			1673	1660	1.1350	1.8702	57.9182	12.0989	$_{v}$ C-N (45)
113			1674	1661	1.1039	1.8218	0.8124	8.8163	$_{v}$ C-N (39)
114			1681	1668	1.9346	3.2223	415.9538	0.9772	$_{v}$ C-N (16)
115			1696	1682	4.2612	7.2250	1340.7862	50.3881	$_{v}$C-N (69)
116			1759	1745	6.8395	12.4785	43.6823	18.6576	$_{υ}$ C-C (66)
117			1778	1764	5.1768	9.6491	22.8958	213.1908	$_{v}$ C-C (78)
118			1783	1769	7.2865	13.6577	79.8400	6.1506	$_{v}$ C-N (45)
127		3066	3289	3087	1.0977	6.9988	33.5926	0.5234	$_{va}$ CH_2 (88)
128			3290	3088	1.0977	7.0050	35.0964	0.5308	$_{va}$ CH_2 (89)
129			3294	3092	1.0994	7.0312	63.4047	0.8658	$_{va}$ CH_2 (89)
130			3295	3093	1.0996	7.0351	32.3292	0.1330	$_{va}$ CH_2 (90)
131			3353	3147	1.0943	7.2499	2.2939	0.1160	$_{va}$ CH_2 (87)
132			3355	3149	1.0934	7.2550	1.8949	0.2065	$_{va}$ CH_2 (88)
133			3356	3150	1.0932	7.2579	1.8665	0.1965	$_{va}$ CH_2 (95)
134			3359	3153	1.0926	7.2676	2.9053	0.1785	$_{va}$ CH_2 (83)
135			3415	3206	1.0912	7.5002	0.1620	0.1405	$_{va}$ C-H (93)
136			3433	3223	1.0915	7.5815	3.4831	0.2776	$_{va}$ C-H (92)
137			3439	3228	1.0944	7.6293	4.9136	0.2704	$_{v}$ C-H (94)
138	3437	3437	3474	3261	1.0929	7.7738	1.8154	3.9170	$_{v}$ C-H (95)

Notes: Scaling factor: 0.992 for frequencies below 1804 cm^{-1} and 0.9387 for frequencies after 1804 cm^{-1} [24].
$_{v}$: Symmetric stretching, $_{va}$: asymmetric stretching, β: in plane bending, γ: out of plane bending, ω: wagging, ρ: rocking, t: torsion, δ: scissoring.

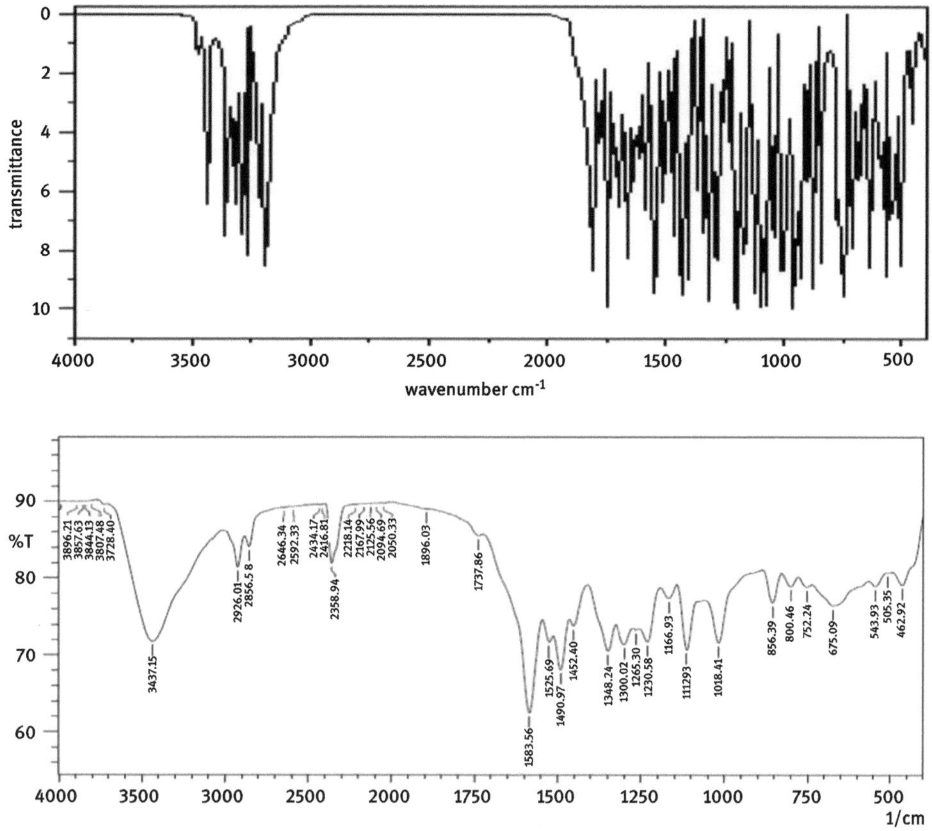

Figure 8.2: (a) Calculated and (b) observed FT-IR spectrum of DMNT.

with the scaled wave number value at 860 and 508 cm^{-1}, respectively, and they are confirmed with the assignments done by Nataraj et al. [24].

8.4.3.3 C-H vibrations

In all aromatic compounds, the band in the region 3100–3000 cm^{-1} [33] corresponds to the stretching of C-H bond. In the present compound DMNT, the stretching vibrations of carbon-hydrogen bond appears at C20-H35, C22-H36, C21-H37 and C23-H38 modes. It is worthwhile to mention that the triazine ring is bridged with two morpholine and electron withdrawing nitrophenoxy rings. Hence the band obtained at 3437 cm^{-1} in both the vibrational spectra is attributed to C-H stretching and the theoretical value obtained is 3261 cm^{-1}. The bands corresponding to both in-plane and out-of-plane vibrations are noted in the region 750–1300 cm^{-1} [34, 35]. The theoretically computed in-plane bending vibration of C-H is obtained at 1459 and 1494 cm^{-1} and it matches with the FT-Raman band at 1463 and 1490 cm^{-1} in

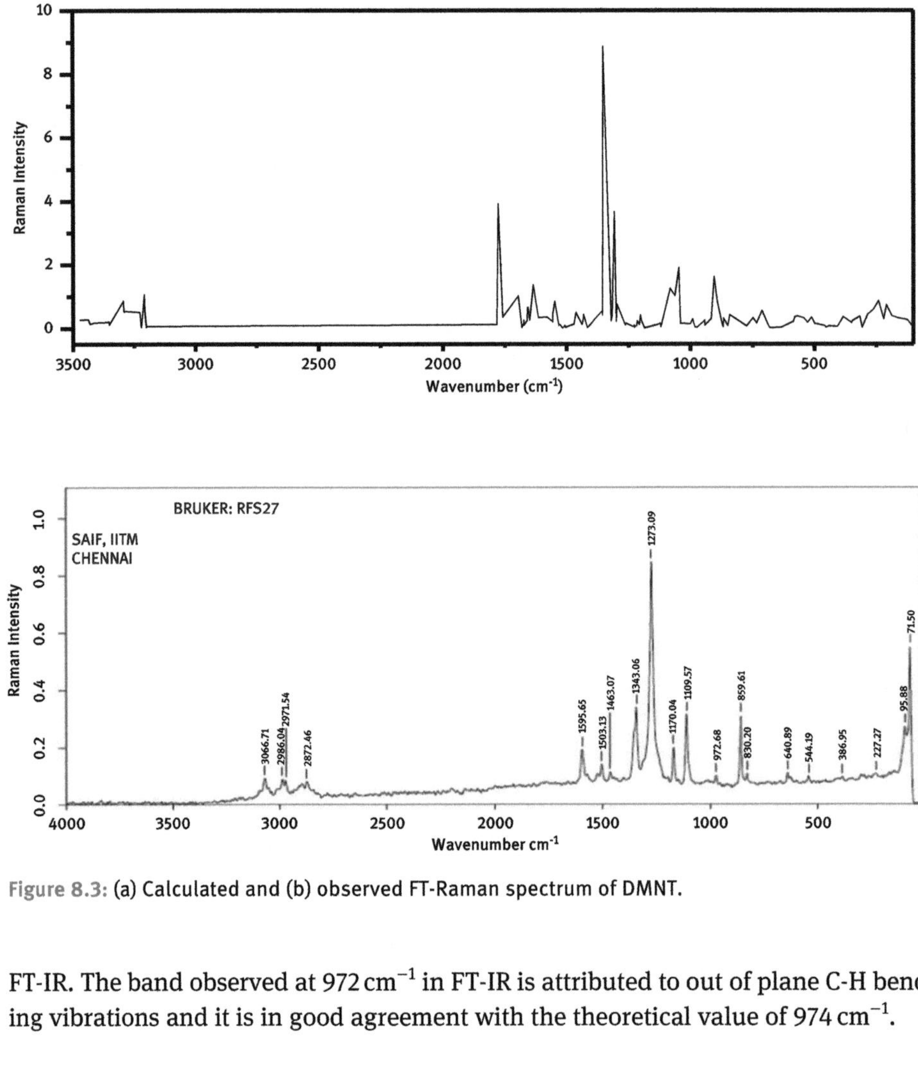

Figure 8.3: (a) Calculated and (b) observed FT-Raman spectrum of DMNT.

FT-IR. The band observed at 972 cm^{-1} in FT-IR is attributed to out of plane C-H bending vibrations and it is in good agreement with the theoretical value of 974 cm^{-1}.

8.4.3.4 Methylene vibrations

The stretching vibrations of methylene group are usually observed in the region 3100–2900 cm^{-1} [36]. For the DMNT, the band at 3066 cm^{-1} in FT-Raman spectra is attributed to the asymmetric stretching vibrations of methylene and it correlates with the theoretical values at 3087–3153 cm^{-1} (mode no: 127–134). FT-Raman band obtained at 2971 and 2986 cm^{-1} corresponds to CH_2 symmetric stretching vibrations and they are in agreement with the scaled frequencies at 3003–3028 cm^{-1} (mode no: 119–126). The scissoring mode of the methylene group usually occurs in the region 1455–1380 cm^{-1} [37]. In the present study for DMNT, the band obtained at 1452 cm^{-1} in FT-IR is due to the scissoring mode. The CH_2 wagging vibrations is notified at 1265 cm^{-1} in FT-IR and the twisting vibrations is obtained at 1112 cm^{-1} in FT-IR and 1109 cm^{-1} in FT-Raman. The band observed at 800 cm^{-1} in FT-IR and 830 cm^{-1} in FT-Raman were due to the rocking vibrations of methylene group. These results correlate well with the scaled frequencies and the literature data [38].

8.4.3.5 Triazine ring vibrations

For DMNT, the band noted at 1503 cm^{-1} in FT-Raman is due to the side chain C-N stretching bands and it agrees with the computed value at 1503 cm^{-1}. The side chain in-plane C-N bending bands and side chain out of plane C-N bending band occur in FT-Raman at 640 and 95 cm^{-1} in FT-Raman, respectively. These vibrations match with the C-N values obtained in melaminium dihydrogen phosphite monohydrate [39].

8.4.3.6 C-C vibrations

The ring C-C stretching vibrations usually appear in the region 1625–1430 cm^{-1} [40, 41]. For DMNT, the band at 1595 cm^{-1} in FT-Raman spectra is assigned to carbon–carbon stretching vibrations. The in plane vibration for carbon for DMNT in FT-IR is observed at 1116 and 1170 cm^{-1} in FT-Raman. The out of plane vibrations in carbon is assigned at 675 cm^{-1} in FT-Raman spectra. These assignments correlate well with the scaled frequencies.

8.4.3.7 C-O-C vibrations

In cyclic ethers, the most significant C-O-C asymmetric stretching vibration occurs in the range of 1150–1085 cm^{-1} [41]. In DMNT, the computed C-O-C asymmetric stretching vibration occurs at 994 cm^{-1} (mode no: 57) and the corresponding experimental FT-IR band appears at 1018 cm^{-1}.

8.4.3.8 Isotopic labelling

The assignments of the normal modes involving C-N, C-C and C-H stretching were verified by substituting ^{13}C, ^{15}N and ^{2}H in both morpholine and phenyl rings. Table SI3 clearly depicts the frequencies which are influenced by the isotopic substitution along with the corresponding isotopic shifts and assignments.

The C-H stretching frequency assignments are confirmed by substituting the hydrogen atoms in morpholine ring and phenyl ring with ^{2}H and the isotopic shifts were computed. The frequencies 3474 and 3433 cm^{-1} are increased by 712 and 722 cm^{-1}, respectively, when all the hydrogen atoms in the morpholine ring are substituted with ^{2}H. The isotopic shifts are observed at 540 and 427 cm^{-1} for the frequencies 3439 and 3415 cm^{-1}, when all the hydrogen atoms in the phenyl ring are substituted with ^{2}H. When each of the carbon atoms in the morpholine ring are replaced with ^{13}C, the frequencies at 1615, 1659, 1660, 1759 and 1778 cm^{-1} are increased by 27, 28, 32, 37 and 39 cm^{-1}, respectively. Thus the C-C assignments are reconfirmed by the isotopic shifts. Likewise, when all the nitrogen atoms in the triazine ring are substituted with ^{15}N, the isotopic shifts observed are 27, 29, 29, 30, 31, 35 cm^{-1} for the modes at 1515, 1649, 1674, 1681, 1696, and 1783 cm^{-1}, respectively. From these isotopic shifts, these frequencies are assigned to C-N stretching vibrations. The N-O stretching is reconfirmed by its isotopic shift with ^{15}N from its original frequency 1566 cm^{-1} by 328 cm^{-1}.

8.4.4 Molecular orbital analysis

HOMO and LUMO are the two important parameters for the investigations of chemical reactions and theoretical study of quantum chemistry [38]. It describes the way by which the molecule interacts with other species and so they are called the frontier orbitals. HOMO specifies the tendency of the orbital to donate electron and the LUMO signifies the ability of the orbital to accept electron. The HOMO–LUMO energy gap (10.67 eV) of DMNT is shown in Figure 8.4.

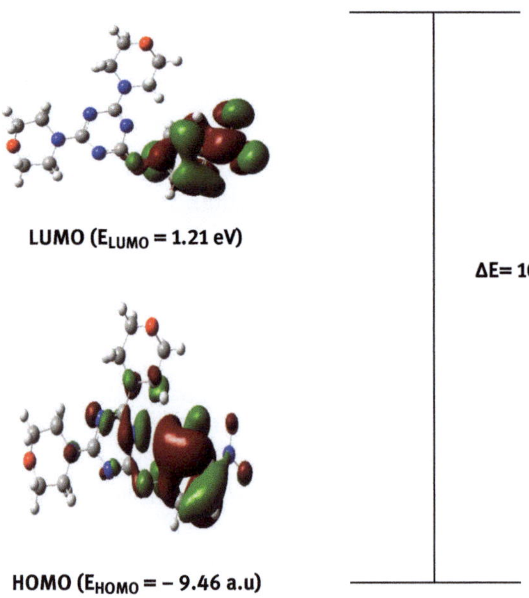

Figure 8.4: The calculated frontiers energies of DMNT.

The HOMO is sited over the nitrogen atoms of the triazine ring and the low energy gap signifies the electron density transfer from the triazine ring to the nitro group and the morpholine ring attached to it. The energy gap identifies the molecular chemical stability, its reactivity and also the kinetic stability [42].

8.4.5 NLO effects

The electrical response of a system can be best studied from its polarizability values. These parameters are used to obtain the NLO properties of the system [43]. The dipole moment (μ) and polarizability (α) were obtained for DMNT and listed in Table SI4 on the basis of the finite-field approach from Gaussian 03 W output which are obtained from the following equations:

$$\alpha_0 = \alpha_{xx} + \alpha_{yy} + \alpha_{zz} \tag{8.1}$$

$$\alpha = 2^{-1/2}\left[(\alpha_{xx} - \alpha_{yy})^2 + (\alpha_{yy} - \alpha_{zz})^2 + (\alpha_{zz} - \alpha_{xx})^2 + 6\alpha_{xx}{}^2\right]^{1/2} \tag{8.2}$$

$$\beta_0 = \left(\beta_x{}^2 + \beta_y^2 + \beta_z^2\right)^{1/2} \tag{8.3}$$

$$\beta_x = \beta_{xxx} + \beta_{xyy} + \beta_{xzz} \tag{8.4}$$

$$\beta_y = \beta_{yyy} + \beta_{xxy} + \beta_{yzz} \tag{8.5}$$

$$\beta_z = \beta_{zzz} + \beta_{xxz} + \beta_{yyz} \tag{8.6}$$

$$\mu = \left(\mu_x^2 + \mu_y^2 + \mu_z^2\right)^{1/2} \tag{8.7}$$

The value of dipole moment and polarizability signifies their NLO property [44]. The theoretical value of dipole moment (μ) was found to be 6.003 D. The magnitude of the molecular hyperpolarizability β decides the magnitude of a NLO system. The first hyperpolarizability value (β) is 3.54897×10^{-30} esu. The dipole moment and first hyperpolarizability is 4.3 and 9.5 times greater than that of urea. Hence DMNT compound is a good candidate of NLO material.

8.4.6 Molecular electrostatic potential (MESP)

The relative polarity of the molecule can be best understood visually from the MESP map. The details about the biological activity and bioactive sites of the molecule can be visualized from the mapping of electron density with MESP. The net electrostatic effect at a point is represented by MESP at that point in the space around a molecule. It also quantifies with dipole moments, electro negativity, partial charges of the molecule. Different colours in the MESP map signify the electrostatic potential at the surface and it increases in the order red < orange < yellow < green < blue. Red regions signifies electrophilic regions, blue regions specifies nucleophilic regions and green region correlates region of zero potential. The MESP map of DMNT shown in Figure 8.5 indicates that the negative regions are present above the oxygen atoms O27 and O28 present in the nitro group and on O25, O11 and O18 atoms and so electrophilic attack can take

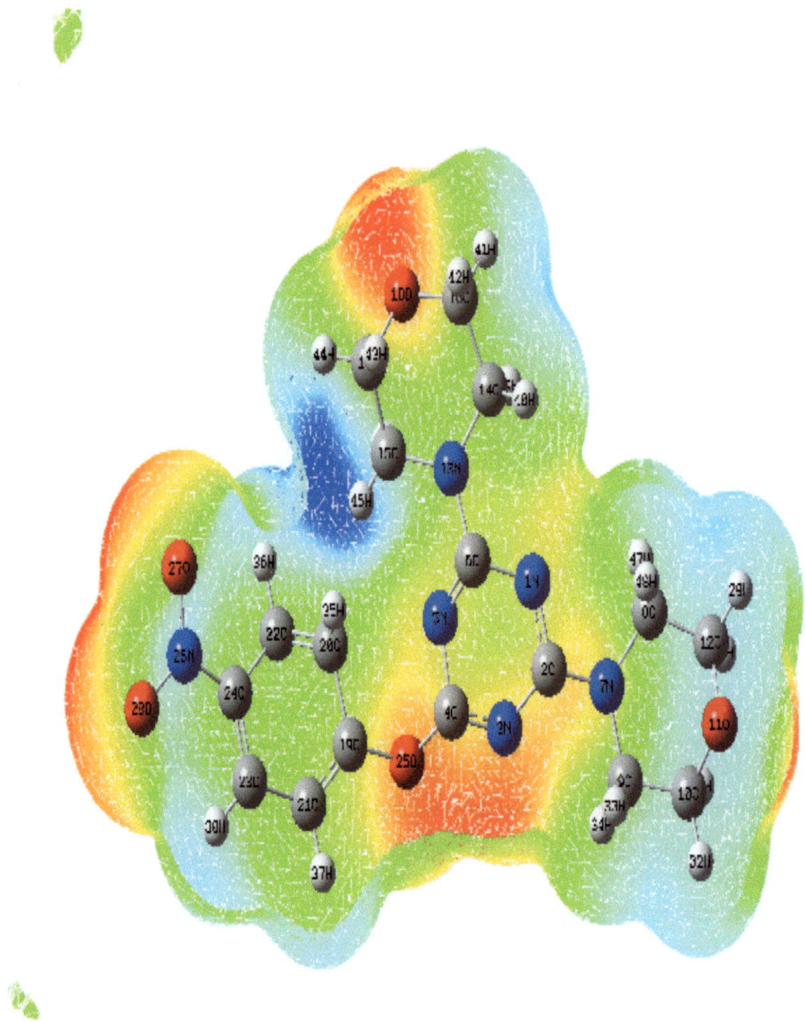

Figure 8.5: The calculated MEP map of DMNT.

place in these sites. The positive regions are situated on the hydrogen atoms of the DMNT molecule; hence nucleophilic attack can take place in these regions. Thus MESP map has been used mainly to analyse reactivity sites and biologically active sites.

8.4.7 Atomic charges

The properties of molecular systems like molecular polarizability, dipole moment, electronic structure depends on atomic charges and they decide the properties of molecular systems. The total charges of atoms of DMNT were calculated by Merz-

Singh-Kollman (MK) scheme and they are given in Table SI5. The Mulliken atomic charges specifies that N1, N3, N5, N7 and N13 have higher negative charges (−0.846e for N1, −0.812e for N3, −0.838e for N5, −0.943e for N7 and −0.943e for N13). The magnitude of the carbon atomic charges ranges from −0.080 to 1.160e and from −0.047e to 0.348e (MK charges). All the hydrogen atoms have a positive charge and nitrogen atoms has a negative charge except N26. The more negative charge on O27 (MK: −0.379e) and O28 (MK: −0.352e) makes N26 more electropositive. The negative charges mainly concentrated on O27 and O28 are involved in charge transfer. Also, N26 (MK: 0.523e) is the most positively charged part and it can undergo nucleophilic attack.

8.4.8 NBO analysis

To specify the charge delocalisation in the bonds and to study the localisation of the lone pairs on oxygen atoms, Natural Bond Analysis has been done for DMNT using NBO 3.1 program. Second order perturbation energy E_2 [45–48] gives description about the non-covalent bonding-antibonding interactions. The NBO results have been listed in Table SI6 and SI7. The intramolecular interaction between lone pair of N7 with antibonding N1-C2 results in a stabilization energy of 41.72 kcal/mol. The most important interaction in the DMNT molecule is between the LP(3)O27 and the antibonding N26-O28. This results in a stabilization energy 161.24 kcal/mol and denotes larger delocalisation. The bond C20-C22 with electron density 1.717 stabilizes the energy of 47.44 kcal/mol to its acceptor antibonding orbital C23-C24. There is a chance of hyperconjugation between N26 and the benzene ring due to the significant decrease of the lone pair orbital occupancy 1.97787 than the other occupancy. The valence hybrid analyses of NBO orbitals proves that electron density distribution around the nitro group is responsible for the polarity of the DMNT compound. The maximum electron density on the oxygen atom is also responsible for the polarity of the molecule. The p-character of oxygen lone pair orbital LP(2) O27 and LP(3) O28 are 99.86 % and 100 %, respectively. This proves that the lone pair orbital involves in the donation of electron in the DMNT compound.

8.4.9 Global reactivity descriptors

Density functional theory helps in the prediction of qualitative chemical parameters such as electronegativity (χ), chemical potential (μ), global hardness (η), global softness (S) and electrophilicity index (ω). Since they are highly useful in analysing the global reactivity they are also called as global reactivity descriptors [49]. They are calculated with Koopman's theorem from the eqs. (8.8)–(8.12) [50–53] and are listed in Table 8.3.

$$\chi = -1/2(\varepsilon_{LUMO} + \varepsilon_{HOMO}) \tag{8.8}$$

$$\mu = -\chi = 1/2(\varepsilon_{LUMO} + \varepsilon_{HOMO}) \tag{8.9}$$

$$\eta = 1/2(\varepsilon_{LUMO} + \varepsilon_{HOMO}) \tag{8.10}$$

$$S = 1/2\eta \tag{8.11}$$

$$\omega = \mu^2/2\eta \tag{8.12}$$

Table 8.3: Chemical hardness, chemical potential of DMNT molecule.

Properties	B3LYP/6-31G(d) values in kcal/mol
HOMO	−913.07
LUMO	116.28
Electronegativity	398.39
Chemical Potential	−398.39
Global Hardness	514.67
Global Softness	6696.60
Electrophilicity index	154.19

Parr et al. [54] reported that electrophilicity index (ω) is a stabilization energy when the compound gains an electron from the surrounding. It is a definite quantity and is similar to the chemical hardness and chemical potential. It is useful in analysing the pollutants based on their reactivity and active sites [55–57]. The chemical potential determines the direction of the charge transfer of the electrons. An electrophile is a chemical moiety capable of gaining electrons from its environment and its energy must decrease after accepting electrons [58]. In the present compound DMNT, the negative values of μ (−398.59 kcal/mol) indicate that charge transfer processes are more significant and it leads to stabilisation through hyperconjugative interactions. The soft molecules are more reactive than the hard molecules because they need small excitation energy. The least value of global hardness (514.67 kcal/mol) and the highest value of global softness (6696.60 kcal/mol) for DMNT molecule indicate their highest inhibition efficiency.

8.4.10 Other molecular properties

The thermodynamic parameters like heat capacity at constant pressure (C_p), entropy (S) and enthalpy changes (ΔH) for DMNT were calculated in the temperature range from 100 to 700 K from the theoretical harmonic frequencies and they listed in Table SI8. As the molecular vibrational intensities increase with temperature [59], these thermodynamic parameters increase with rise of temperature. The correlation graphs for the temperature and properties are shown in Fig. SI1. These properties helpful information to predict the directions of chemical reactions in accordance with the law of thermodynamics [60–62].

8.4.11 UV spectrum and electronic properties

In order to understand the electronic transitions in the molecule, time independent TD-DFT calculations involving singly excited electronic states was performed. The calculated excitation energies, oscillator strength (*f*), wavelength of absorption (λ) and spectral assignments are given in Table 8.4. Ethanol was used as a solvent to simulate the electronic absorption. Figure 8.6(b) represents the computed electronic spectra of DMNT compound. The electronic spectra are recorded for DMNT within 200–800 nm and is shown in Figure 8.6(a). DMNT has oxygen and nitrogen with lone pair of electrons and with this $\pi \rightarrow \pi^*$ transition is possible. The strong transition observed at 3.80 eV (326.4 nm) and at 3.52 eV (301.5 nm) are assigned to $\pi \rightarrow \pi^*$.

Table 8.4: Theoretical electronic absorption spectra of DMNT using TD-B3LYP/6-31G(d) method.

Ethanol			Gas			Assignments
λ(nm)	E (eV)	f	λ(nm)	E (eV)	f	
403.26	3.0745	0.0012	397.5	3.0652	0.0014	σ — σ*
326.40	3.7986	0.2017	301.5	3.5232	0.2865	π → π*
309.63	4.0042	0.1316	272.5	3.9989	0.1592	n — π*

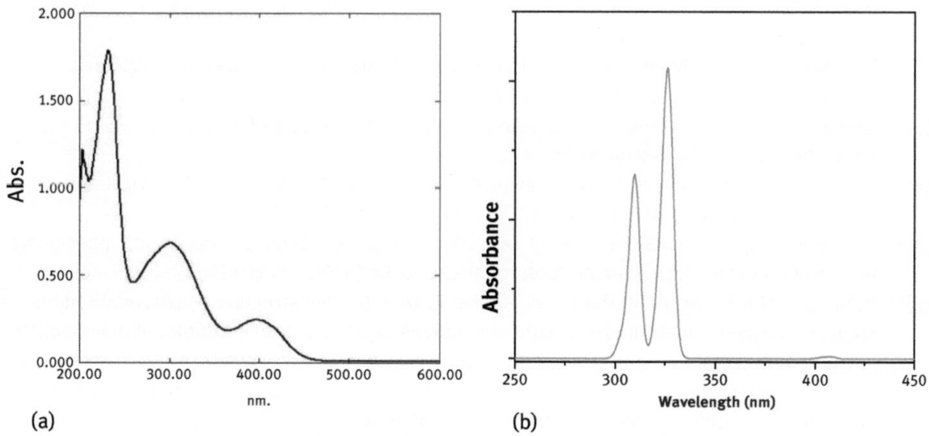

Figure 8.6: (a) Experimental and (b) Calculated UV-Vis spectra of DMNT.

The intense band noted at 301.5 nm can be attributed to resonating features of π-electrons in the triazine and morpholine rings. The medium transition observed at 272.5 nm is attributed to n — π* transition. The low energy absorption peak found at 397.5 nm belongs to the dipole-allowed σ — σ* from HOMO to LUMO. It is also due to the excitation of sigma electrons which are localized on the entire molecule.

8.5 Conclusions

In this study, DMNT is synthesized and computations are done based on DFT method. The vibrational spectra, FT-IR and FT-Raman, were recorded and the vibrational frequencies analysis agrees well with the experimental results and isotopic shifts. The calculated geometrical parameters are in good agreement with analogues of DMNT. HOMO–LUMO energy gap explains that significant charge transfer interactions have taken place within DMNT. The calculated dipole moment and first order hyperpolarizability results indicate that DMNT has NLO properties. NBO analysis reveals the hyperconjugative interaction taking place within the molecule. The MEP shows the electrophilic and nucleophilic sites of DMNT compound. UV-Vis spectra reveal the types of electronic transitions taking place within DMNT. The findings of this research work add to the literature of DMNT for its potential uses and applications.

Acknowledgements: The authors acknowledge the facilities from their respective Universities.

References

[1] Elyakov GB, Stonik VA, Makareva TN. Heterocyclic compounds of marine organisms (Review). Chem Heterocycl Compd. 1977;13:345–59.
[2] Newman DJ, Cragg GM. Marine natural products and related compounds in clinical and advanced preclinical trials. J Nat Prod. 2004;67:1216–38.
[3] Roje S. Bioconversion process for producing nylon-7, nylon-7,7 and polyesters. Photochemistry. 2007;68:1904–21.
[4] Jain KS, Chitre TS, Miniyar PB, Kathiravan MK, Bendre VS, Veer VS, et al. Biological and medicinal significance of pyrimidines. Curr Sci. 2006;90:793–803.
[5] Leiningera E, Belousov AB. Recovery of network-driven glutamatergic activity in rat hippocampal neurons during chronic glutamate receptor blockade. Brain Res. 2009;1251:87–102.
[6] Ojha H, Gahlot P, Tiwari AK, Pathak M, Kakkar R. Quantitative structure activityrelationship study of 2, 4, 6-trisubstituted-s-triazine derivatives as antimalarial inhibitors of plasmodium falciparum dihydrofolate reductase. Chem Biol Drug Des. 2011;77:57–62.
[7] Alekseeva NV, Yakhontov LN. Reactions of pyridines, pyrimidines and 1,3,5-triazines with nucleophilic reagents. Russ Chem Rev. 1990;59:514–30.
[8] Charushin VN, Chupakhin ON, Van der Plas HC. Reactions of azines with bifunctional nucleophiles: cyclizations and ring transformations. Adv Heterocycl Chem. 1988;43:301–53.

[9] Loos R, Niessner R. Analysis of atrazine, terbutylazine and their n-dealkylated chloro and hydroxy metabolites by solid-phase extraction and gas chromatography-mass spectrometry and capillary electrophoresis- ultraviolet detection. J Chromatograpr A. 1999;835:217–29.
[10] Vuddhakul V, Jacobsen NW, Rose SE, Joanoni B, Seow WK, Thong YH. 1-methyl-3-phenyl-1,2,4-triazinium-5-olate: a new zwitterion with cytotoxic activity against human cancer cell-lines. Cancer Lett. 1988;42:29–35.
[11] Hirai K, Sugimoto H, Mizushima T. Jpn Kokai Tokkyo Koho 1986. JP 61134389 A2.
[12] Bierowska-Charytonowicz D, Konieczny M. Rocz Chem. 1973;47:2199.
[13] Oettmeier W, Hilip U, Draper W, Fedtke C, Schmidt RR. Pestic Sci. 1991;33:399.
[14] Kranz E, Santel HJ, Luerssen K, Schmidt RR, Krauskopf B. Ger Offen 1990. DE 3917043 A1.
[15] Boehner B, Tobler H. Eur Pat Appl 1985. EP 150677.
[16] Janczak J, Perpetuo GJ. Melaminium glutarate monohydrate. Acta Crystallogr Sect C. 2001;57:1431–3.
[17] Marchewka MK. Infrared and Raman spectra of melaminium chloride hemihydrate. Mat Sci Eng B. 2002;95:214–21.
[18] Frisch MJ, Trucks GW, Schlegal HB, Scuseria GE, Robb MA, Cheesman JR, et al. GAUSSIAN 03, Revision A.02. Pittsburg, PA: Gaussian, Inc., 2003.
[19] Frisch A, Dennington R, Keith T, Milliam J, Nielsen AB, Holder AJ, et al. Gaussview reference, Version 4.0. Pittsburgh: Gaussian Inc., 2007.
[20] Vennila JP, John Thiruvadigal D, Kavitha HP, Chakkaravathi G, Manivannan V. N-[-2,(3,4-dimethoxyphenyl)ethyl]-nmethylnaphthalene-1-sulfonamide. Acta Cryst E. 2011;67:o2451.
[21] Arivazhagan M, Meenakshi R. Quantum chemical studies on structure of 1-3-dibromo-5-chlorobenzene. Spectrochim Acta A. 2011;82:316–26.
[22] Fogarasi G, Pulay P. Vibrational spectra and structure, vol. 14. Amsterdam: Elsevier, 1985:125–219.
[23] Pulay P. In: Schalfer HF, editor. Applications of electronic structure theory, modern theoritical chemistry, vol. 4. New York: Springer, 1997:153.
[24] Nataraj A, Balachandran V, Karthick T. FTIR and Raman spectra, DFT and SQMFF calculations for geometrical interpretation and vibrational analysis of 3-nitro-p-toluic acid. J Mol Struct. 2012;1022:94–108.
[25] Polavarapu PL. Abinitio vibrational Raman and Raman optical activity spectra. J Phys Chem. 1990;94:8106–12.
[26] Keresztury G, Holly S, Varga J, Besenyei GA, Wang Y, Durig JR. Electronic absorption and vibrational spectra and nonlinear optical properties of 4-methoxy-2-nitroaniline. Spectrochim Acta Part. 1993;49A:2007-26.
[27] Keresztury G. Raman spectroscopy: theory. In: Chalmers JM, Griffiths PR, editors. Hand book of vibrational spectroscopy, vol. 1. Chicester: Wiley, 2002.
[28] Jamroz MH. Vibrational energy distribution analysis: VEDA 4 Program Warsao. Poland, 2004.
[29] Meishlich EK, Meislich H, Sharefkin J. 3000 solved problems in organic chemistry, vol. 2. New York: McGraw-Hill, 1993.
[30] George S. Infrared and Raman characteristic group frequencies-tables and charts, 3rd ed. New York: Wiley, 1993.
[31] Kovacs A, Keresztury G, Izvekow V. Intramolecular hydrogen bonding in 2-nitrorezorcinol. A combined FT-IR, FT-Raman and computational study. Chem Phys. 2000;253:193–204.
[32] Rao JM. Organic spectroscopy, principles and applications. New Delhi: Narosa Publishing House, 2000.
[33] Varsanyi G. Vibrational spectra of benzene derivatives. New York: Academic Press, 1969.
[34] Socrates G. Infrared characteristic group frequencies. New York: Wiley, 1980.

[35] Karakurt T, Dincer M, Cukurovali A, Yilmaz I. Ab initio and semi-empirical computational studies on 5-hydroxy4-methyl-5, 6-di-pyridin-2-yl-4, 5-dihydro-2H-[1, 2, 4] triazine-3-thione. J Mol Struct. 2011;991:186–201.
[36] Silverstein M, Clyyon Basseler G, Morill C. Spectrometric I identification of organic compounds. New York: Wiley, 1981.
[37] Lin-Vein D, Colthup NB, Fateley WG, Grasseli JG. The hand book of Infrared and Raman characteristic frequencies of organic molecules. San Diego: Academic Press, 1991.
[38] Subhash Chandrabose S, Saleem H, Erdogdu Y, Rajarajan G, Thanikachalam V. FT-Raman, FT-IR spectra and total energy distribution of 3-pentyl-2,6-diphenylpiperidin-4-one: DFT method. Spectrochim Acta A. 2011;82:260–9.
[39] Arjunan V, Kalaivani M, Marchewka MK, Mohan S. Crystal structure, vibrational and DFT simulation studies of melaminium dihydrogen phosphite monohydrate. J Mol Struct. 2013;1045:160–70.
[40] Varsanyi G. Assignments of vibrational spectra of seven hundred benzene derivatives, vol. 1-2. London: Adam Hilger, 1974.
[41] Bellanmy LJ. The infrared spectra of complex molecules. London: Chapman and Hall, 1980.
[42] Balachandran V, Janaki A, Nataraj A. Theoretical investigations on molecular structure, vibrational spectra, HOMO, LUMO, NBO analysis and hyperpolarizability calculations of thiophene-2-carbohydrazide. Spectrochim Acta A. 2014;118:321–30.
[43] Politzer P, Laurence PR, JayaSuriya K, McKinney J. Molecular electrostatic potentials: an effective tool for the elucidation of biochemical phenomena. Environ Health Perspect. 1985;61:191–202.
[44] Pegu D. Int J Sci Res. 2012;469–74. ISSN: 2319–7064.
[45] Reed AE, Weinhold F. Natural localized molecular orbitals. J Chem Phys. 1985;83:1736–40.
[46] Reed AE, Weinhold RB, Weinhold F. Natural population analysis. J Chem Phys. 1985;83:735–46.
[47] Reed AE, Wienhold F. Natural bond orbital analysis of near-hartree-fock water dimer. J Chem Phys. 1983;78:4066–73.
[48] Foster JP, Weinhold F. Natural hybrid orbitals. J Am Chem Soc. 1980;102:7211–8.
[49] Parthasarathi R, Padmanabhan J, Subramanian V, Maiti B, Chattraj PK. Chemical reactivity profiles of two-selected poly chlorinated biphenyls. J Phys Chem A. 2008;107:10346–52.
[50] Thanikaivelan P, Subramanian V, Raghava Rao J, Nair BV. Application of quantum chemical descriptor in quantitative structure activity and structure property relationship. Chem Phys Lett. 2000;323:59–70.
[51] Parthasarathi R, Padmanabhan J, Elango M, Subramanian V, Chattraj PK. Intermolecular reactivity through the generalized electrophilicity concept. Chem Phys Lett. 2004;394:225–30.
[52] Parthasarthi R, Padmanabhan J, Subramanian V, Maiti B, Chattraj PK. Toxicity analysis of 33'44'5-pentachloro biphenyl through chemical reactivity and selectivity profiles. Curr Sci. 2004;86:535–42.
[53] Parthasarthi R, Padmanabhan J, Subramanian V, Sarkar U, Maiti B, Chattraj PK. Toxicity analysis of benzidine through chemical reactivity and selectivity profiles: a DFT approach. Internet Electron J Mol Des. 2003;2:798.
[54] Parr RG, Yang W. Density functional theory of atoms and molecules. Oxford, New York: Oxford University Press, 1989.
[55] Pearson RG. Absolute electronegativity and hardness: applications to organic chemistry. J Org Chem. 1989;54:1430–2.
[56] Parr RG, Pearson RG. Absolute hardness: companion parameter to absolute electronegativity. J Am Chem Soc. 1983;105:7512–6.
[57] Geerlings P, De Proft F, Langenaekar W. Conceptual density functional theory. Chem Rev. 2003;103:1793–873.

[58] Singh RN, Kumar A, Tiwari RK, Rawat P. A combined experimental and theoretical (DFT and AIM) studies on synthesis, molecular structure, spectroscopic properties and multiple interactions analysis in a novel ethyl-4-[2-(thiocarbamoyl) hydrazinylidene]-3, 5-dimethyl-1h-pyrrole-2- carboxylate and its dimer. Spectrochim Acta. 2013;112:182–90.
[59] Ott JB, Boerio-Goates J. Calculations from statistical thermodynamics. USA: Academic Press, 2000.
[60] Zhang R, Dub B, Sun G, Sun Y. Molecular modeling and spectroscopic studies of benzothiazole. Spectrochim Acta. 2010;75A:115–1124.
[61] Kavipriya R, Kavitha HP, Karthikeyan B, Natraj A. Molecular structure, spectroscopic (FT-IR, FT-Raman), NBO analysis of N,N'Diphenyl-6-piperidinyl-1-yl-[1,3,5]-triazine-2,4-diamine. Spectrochim Acta A. 2015;156:476–87.
[62] Luque FJ, Lopez JM, Orozco M. Perspective on Electrostatic Interactions of a solute with a continuum. A direct utilization of ab initio molecular potentials for the prevision of solvent effects. Theor Chem Acc. 2000;103:343–5.

Supplementary Material

The online version of this article offers supplementary material (https://doi.org/10.1515/psr-2019-0003).

R. Kavipriya, Helen P. Kavitha, Jasmine P. Vennila, Lydia Rhyman and Ponnadurai Ramasami

9 Spectroscopic and DFT studies of 2,4-dichloro-*N*-phenethylbenzenesulfonamide

Abstract: The Fourier-transform infrared spectroscopy (FT-IR) spectrum of 2,4-dichloro-*N*-phenethylbenzenesulfonamide (DPBS) was obtained and the compound was studied theoretically. The optimized geometry, total electronic energy and vibrational wavenumbers of DPBS were examined using Hartree–Fock (HF) and density functional theory (DFT) method such as B3LYP, BP86 and M06 functionals with the basis set of 6-311++G(d,p) for all atoms. A complete vibrational assignment was studied for DPBS. The molecular orbital energies, polarizability and thermodynamic properties of DPBS were also computed. Analysis of molecular orbitals reveals the parameters such as chemical potential, chemical hardness and electrophilicity index. The molecular properties such as electric dipole moment μ, polarizability α, and hyperpolarizability β reveal the non-linear optical (NLO) property of DPBS. Natural bond analysis study reveals charge delocalization of the molecule. The experimental and computational results are found to have good agreement among themselves. The results of this work will pave the way for further insight in the study of the applications of DPBS.

Keywords: 24-dichloro-N-phenethylbenzenesulfonamide (DPBS), spectroscopy, strucuture

---

This article has previously been published in the journal Physical Sciences Reviews. Please cite as: Kavipriya, R., Kavitha, H. P., Vennila, J. P., Rhyman, L., Ramasami, P. Spectroscopic and DFT studies of 2,4-dichloro-N-phenethylbenzenesulfonamide *Physical Sciences Reviews* [Online] 2020, 5 DOI: 10.1515/psr-2019-0002.

https://doi.org/10.1515/9783110631623-009

Graphical Abstract:

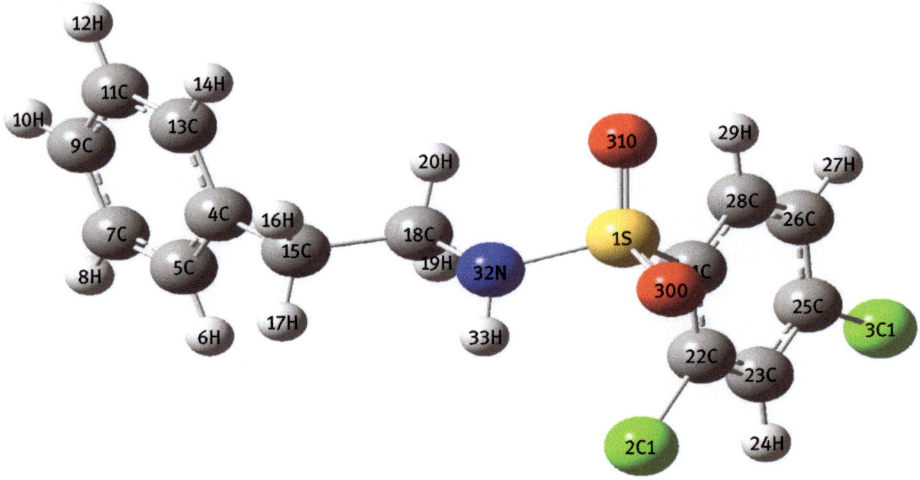

9.1 Introduction

Sulfonamide drug is an important member of synthetic anti-bacterial drugs containing hetero atoms such as sulfur and nitrogen atoms. They are important biologically active compounds [1]. The infections that occur in the central nervous system, respiratory, gastrointestinal and urinary tract can be healed by sulphonamide drugs. They are anti-bacterials and are widely used in the treatment of domestic animals due to their low cost and simple way of administration. Heteroaromatic compounds containing sulfonamides and their derivatives play a vital role for the synthesis of pharmacological agents [2]. The anti-bacterial sulfonamides are used in the treatment of cancer [3, 4]. Sulfonamides have medicinal applications such as diuretic [5], anti-epileptic [6], hypoglycemic [7], anti-tumor [8], anti-microbial [9], anti-cancer [10], anti–inflammatory [11] and anti–viral agents as well as HIV protease inhibitors drugs [12]. The geometrical parameters of 2,4-dichloro-*N*-phenethylbenzenesulfonamide (DPBS) from its crystal were determined by Suneel et al. [13]. Despite the importance and applications of DPBS, to the best of our knowledge, there has not been any detailed study to interpret the vibrational spectra and other molecular parameters of the compound. These have driven us towards the study of the molecular structure of DPBS and their vibrational spectral analysis. The molecular properties, thermodynamic properties, polarizability and non-linear optical (NLO) were also studied. A natural bonding orbital (NBO) analysis of DPBS was also carried out.

9.2 Experimental details

The pure grade DPBS was purchased from Sigma-Aldrich Chemicals, USA. The Fourier-transform infrared spectrospcoy (FT-IR) spectrum of the compound was recorded in Bruker IFS 66 V spectrometer in the range of 4000–100 cm^{-1}.

9.3 Computational methods

All calculations were computed using GAUSSIAN program [14]. The optimized structure of DPBS is obtained using HF and DFT methods such as B3LYP, BP86 [15–18] and M06 [19] with 6-311++G(d,p) basis set. The vibrational assignments were done by animation of modes in GAUSSVIEW program [14] and from normal coordinate analysis. The molecular orbital energies reveals the properties such as chemical potential, chemical hardness and electrophilicity indices. The NBO calculations predict the possibility of the intermolecular hydrogen bonding, hyperconjugative interactions and delocalization of electron density.

9.4 Results and discussion

9.4.1 Molecular geometry

The DPBS molecule contains two phenyl rings connected with two methylene, –NH and $-SO_2$ groups. The optimized geometrical parameters of DPBS are presented in Table 9.1. The position of the atoms with numbers is depicted in Figure 9.1. It is noticed that theoretical bond lengths are slightly greater than the experimental values since theoretical calculations are assumed to have isolated molecules in the gaseous phase and the experimental results are associated with solid state. Comparing the geometrical parameters obtained from the four theoretical methods, it is found that the HF values correlate well with the experimental values. It is also observed that the values of BP86 show greater deviations when compared to the other methods. The largest deviation in bond length was found to be 0.138, 0.155, 0.166 and 0.157 Å and the smallest deviation in bond length was 0.001, 0.003, 0.004, and 0.012 Å using the HF method and the B3LYP, BP86 and M06, respectively. From Table 9.1, it can be observed that significant differences (8°) are found in the calculated bond angle for S1-N32-H33 when compared to their experimental value. The other bond angles of DPBS are found to be matching with the experimental values.

Table 9.1: Optimized structural parameters of DPBS.

Parameters	HF	B3LYP	BP86	M06	XRD
Bond length (Å)					
S1–C21	1.790	1.821	1.831	1.801	1.777
S1–O30	1.418	1.455	1.470	1.443	1.423
S1–O31	1.422	1.458	1.474	1.446	1.429
S1–N32	1.619	1.666	1.689	1.657	1.600
Cl2–C22	1.738	1.753	1.757	1.742	1.730
Cl3–C25	1.736	1.750	1.753	1.737	1.730
C4–C15	1.513	1.512	1.516	1.500	1.512
C5–H6	1.076	1.085	1.095	1.088	0.930
C5–C7	1.351	1.393	1.401	1.388	1.385
C15–C18	1.529	1.536	1.542	1.521	1.499
C18–N32	1.462	1.471	1.477	1.459	1.470
C21–C22	1.391	1.398	1.406	1.393	1.388
N32–H33	0.997	1.0148	1.0256	1.016	0.860
Bond angle (°)					
C21–S1–O30	108.3	108.3	109.1	109.1	108.3
C21–S1–O31	105.1	105.1	105.1	105.1	106.6
C21–S1–N32	107.6	107.6	106.1	106.1	107.4
O30–S1–O31	120.9	121.9	121.9	121.9	119.0
O30–S1–N32	106.7	106.3	106.4	106.4	107.8
O31–S1–N32	107.4	106.8	107.1	107.1	107.6
C5–C4–C13	118.4	118.4	118.7	118.7	120.6
C13–C4–C15	120.5	120.6	120.3	120.3	120.0
C15–C18–N32	109.6	109.8	110.1	110.1	110.4
H16–C15–H17	107.2	107.1	107.1	107.1	107.8
H19–C18–H20	108.1	108.1	107.5	107.5	108.1
S1–N32–C18	121.6	120.3	118.8	118.8	120.2
S1–N32–H33	111.7	110.2	110.1	110.1	119.9

9.4.2 Normal coordinate analysis

There are 93 fundamental modes of vibration for DPBS molecule. Among the 93 vibrations, the numbers of in-plane and out-of-plane vibrations are 65 and 28, respectively. The normal coordinate analysis gives a elaborate explanation about vibrational modes. The symmetry coordinates are obtained for DPBS from the set of internal coordinates. The entire set of 87 internal coordinates (with 22 redundancies) were denoted as mentioned in Table 9.2. Also a non-redundant set of local symmetry coordinates were described by linear combinations of internal coordinates [20–23] and they are depicted in Table 9.3.

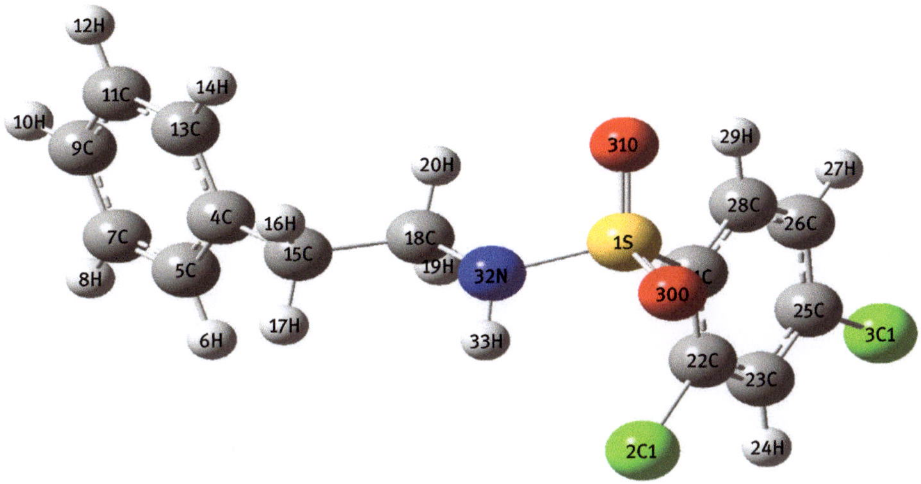

Figure 9.1: Optimized molecular structure of DPBS.

Table 9.2: Definition of internal coordinates associated with DPBS.

No(i)	Symbol	Type	Definition
Stretching			
1–12	P_i	C–C (aromatic)	C9–C11, C11–C4, C4–C13, C13–C5, C5–C7, C7–C9, C26–C25, C25–C23, C23–C22, C22–C21, C21–C28, C28–C26
13–14	P_i	C–C	C14–C15, C15–C18
15–22	r_i	C–H	C9–H10, C11–H12, C13–H14, C5–H6, C7–H8, C26–H27,C23–H24, C28–H29
23–26	r_i	C–H (ethylene)	C15–H16, C15–H17, C18–H19, C18–H20
27	S_i	N–H	N32–H33
28	q_i	C–N	C18–N32
29	ϕ_i	N–S	N32–S1
30	Q_i	C–S	S1–C21
31–32	R_i	C–Cl	C25–Cl3, C22–Cl2
34	γ_i	S–O	S1–O31, S1–O30
In plane bending			
35–40	βi	t (ring 1)	C9–C11–C4, C11–C4–C13, C4–C13–C5, C13–C5–C7, C5–C7–C9, C7–C9–C11
41–46	βi	t (ring 2)	C26–C25–C23, C25–C23–C22, C23–C22–C21, C22–C21–C28, C21–C28–C26, C28–C26–C25

Table 9.2 (continued)

No(i)	Symbol	Type	Definition
47–56	βi	C–C–H	C9–C11–H12, C4–C11–H12, C4–C13–H14, C5–C13–H14, C13–C5–H6, C7–C5–H6, C5–C7–H8, C9–C7–H8, C7–C9–H10, C11–C9–H10
57–62	βi	C–C–H	C28–C26–H27, C25–C26–H27, C25–C23–H24, C22–C23–H24, C21–C28–H29, C26–C28–H29
63–66	σi	C–C–Cl	C26–C25–Cl3, C23–C25–Cl3, C23–C22–Cl2, C23–C25–Cl2
67–70	βi	C–C–H	C4–C15–H16, C4–C15–H17, C15–C18–H19, C15–C18–H20
71–72	βi	H–C–H	H6–C5–H17, H19–C18–H20
73–74	η_i	C–C–S	C28–C21–S1, C22–C21–S1
75–76	Σ_i	C–S–O	C21–S1–O30, C21–S1–O31
77	Σ_i	O–S–O	O30–S1–O31
78	πi	C–C–N	C15–C18–N32
79	πi	H–C–N	H20–C18–N32
80–82	βi	C–C–C	C4–C15–C18, C11–C4–C15, C13–C14–C15
83	βi	N–C–C	N32–C15–C18
84	η_i	C–N–S	C18–N32–H33
85	η_i	N–S–C	N32–S1–C21
86	Φi	C–N–H	C18–N32–H33
87	Φ_i	S–N–H	S1–N32–H33
Out of plane bending			
88–93	τi	t (Ring 1)	C9–C11–C4–C13, C11–C4–C13–C5,C4–C13–C5–C7, C13–C5–C7–C9,C5–C7–C9–C11, C7–C9–C11–C4
94–99	τi	t (Ring 2)	C26–C25–C23–C22, C25–C23–C22–C21, C23–C22–C21–C28, C22–C21–C28–C26, C21–C28–C26–C25, C28–C26–C25–C23
100–104	ωi	C–H (aromatic)	H10–C9–C7–C11, H12–C11–C9–C4, H14–C13–C5–C4, H6–C5–C7–C13, H8–C7–C9–C5
105–107	ωi	C–H (aromatic)	H27–C26–C28–C25, H24–C23–C22–C25, H29–C28–C21–C26
108–109	ωi	C–Cl	Cl3–C25–C26–C23, Cl2–C22–C21–C23
110–111	ωi	CH_2 wagging	H19–C18–C15–H16, H20–C18–C15–H17
112	ωi	SO_2 wagging	O31–S1–N32–C21
113–114	ωi	C–C–C–H	C4–C15–C18–H19, C4–C15–C18–H20
115	ωi	H–C–C–S	H29–C28–C21–S1
116	ωi	C–C–N–S	C15–C18–N32–S1
117	ωi	C–C–N–H	C15–C18–N32–H33
118	ωi	H–C–N–H	H19–C18–N32–H33

Table 9.3: Definition of local symmetry coordinates for DPBS.

No(i)	Type	Definition
Stretching		
1–12	C–C (ar)	$P_1, P_2, P_3, P_4, P_5, P_6, P_7, P_8, P_9, P_{10}, P_{11}, P_{12}$
13–14	C–C	P_{13}, P_{14}
15–22	C–H	$r_{15}, r_{16}, r_{17}, r_{18}, r_{19}, r_{20}, r_{21}, r_{22}$
23	$(CH_2)_1$ss	$(r_{23}+r_{24})/\sqrt{2}$
24	$(CH_2)_1$ as	$(r_{23}-r_{24})/\sqrt{2}$
25	$(CH_2)_2$ss	$(r_{25}+r_{26})/\sqrt{2}$
26	$(CH_2)_2$ as	$(r_{25}-r_{26})/\sqrt{2}$
27	N–H	S_{27}
28	C–N	q_{28}
29	N–S	ϕ_{29}
30	C–S	Q_{30}
31–32	C–Cl	R_{31}, R_{32}
33	SO_2ss	$(\gamma_{33}+r_{34})/\sqrt{2}$
34	SO_2 as	$(\gamma_{33}-r_{34})/\sqrt{2}$
In plane bending		
35	R1 trigd	$(\beta_{35}-\beta_{36}+\beta_{37}-\beta_{38}+\beta_{39}-\beta_{40})/\sqrt{6}$
36	R1 symd	$(-\beta_{35}-\beta_{36}+2\beta_{37}-\beta_{38}-\beta_{39}-2\beta_{40})/\sqrt{12}$
37	R1 asymd	$(\beta_{35}-\beta_{36}+\beta_{38}-\beta_{39})/2$
38	R2 trigd	$(\beta_{41}-\beta_{42}+\beta_{43}-\beta_{44}+\beta_{45}-\beta_{46})/\sqrt{6}$
39	R2 symd	$(-\beta_{41}-\beta_{42}+2\beta_{43}-\beta_{44}-\beta_{45}-2\beta_{46})/\sqrt{12}$
40	R2 asymd	$(\beta_{41}-\beta_{42}+\beta_{44}-\beta_{45})/2$
41–45	bCCH	$(\beta_{47}-\beta_{48})/\sqrt{2}, (\beta_{49}-\beta_{50})/\sqrt{2}, (\beta_{51}-\beta_{52})/\sqrt{2}, (\beta_{53}-\beta_{54})/\sqrt{2}, (\beta_{55}-\beta_{56})/\sqrt{2}$
46–48	bCCH	$(\beta_{57}-\beta_{58})/\sqrt{2}, (\beta_{59}-\beta_{60})/\sqrt{2}, (\beta_{61}-\beta_{62})/\sqrt{2}$
49–50	bCCl	$(\sigma_{63}-\sigma_{64})/\sqrt{2}, (\sigma_{65}-\sigma_{66})/\sqrt{2}$
51	CH_2 scissoring	$2(\Psi_{67}-\Psi_{68}-\Psi_{69})/\sqrt{6}$
52	CH_2 scissoring	$2(\Psi_{71}-\Psi_{72}-\Psi_{70})/\sqrt{6}$
53	CH_2 rocking	$(\Psi_{68}-\Psi_{69})/\sqrt{2}$
54	CH_2 rocking	$(\Psi_{72}-\Psi_{70})/\sqrt{2}$
55, 56	CH_2 twisting	$(\Psi_{68}+\Psi_{69})/\sqrt{2}, (\Psi_{72}+\Psi_{70})/\sqrt{2}$
57	Bccs	$(\eta_{73}-\eta_{74})/\sqrt{2}$
58	SO_2 rocking	$(\sum_{75}-\sum_{76})/\sqrt{2}$
59	SO_2 twisting	$(\sum_{75}+\sum_{76})/\sqrt{2}$
60	SO_2 scissoring	$(2\sum_{77}-\sum_{75}+\sum_{76})/\sqrt{6}$
61	bCN	$(\pi_{78}-\pi_{79})/\sqrt{2}$
62–63	bCCC	$(\beta_{80}-\beta_{81})/\sqrt{2}, (\beta_{82}-\beta_{83})/\sqrt{2}$
64	bNS	$(\eta_{84}-\eta_{85})/\sqrt{2}$
65	bNH	$(\Phi_{86}-\Phi_{87})/\sqrt{2}$
Out of plane bending		
66	t R1 trigd	$(\tau_{88}-\tau_{89}+\tau_{90}-\tau_{91}+\tau_{92}-\tau_{93})/\sqrt{6}$
67	t R1 symd	$(\tau_{88}-\tau_{90}+\tau_{91}-\tau_{93})/\sqrt{2}$
68	t R1 asymd	$(-\tau_{88}+2\tau_{89}-\tau_{90}-\tau_{91}+2\tau_{92}-\tau_{93})/\sqrt{12}$
69	t R2 trigd	$(\tau_{94}-\tau_{95}+\tau_{96}-\tau_{97}+\tau_{98}-\tau_{99})/\sqrt{6}$

Table 9.3 (continued)

No(i)	Type	Definition
70	t R2 symd	$(\tau_{94}-\tau_{96}+\tau_{97}-\tau_{99})/\sqrt{2}$
71	t R2 asymd	$(-\tau_{94}+2\tau_{95}-\tau_{96}-\tau_{97}+2\tau_{98}-\tau_{99})/\sqrt{12}$
72–76	ωCH_{ar}	$\omega_{100}, \omega_{101}, \omega_{102}, \omega_{103}, \omega_{104}$
77–79	ωCH_{ar}	$\omega_{105}, \omega_{106}, \omega_{107}$
80–81	ωCCl	$\omega_{108}, \omega_{109}$
82–83	CH_2 wagging	$\omega_{110}, \omega_{111}$
84	SO_2 wagging	ω_{112}
85–86	ωCC	$\omega_{113}, \omega_{114}$
87	ωNS	ω_{115}
88	ωCS	ω_{116}
89–90	ωNH	$\omega_{117}, \omega_{118}$
91	ωCCl	ω_{119}
92	ωCS	ω_{120}
93	ωNS	ω_{121}

9.4.3 Vibrational frequencies

There are 33 atoms in the DPBS molecule and hence it has 93 modes of vibrations. The experimentally observed and theoretically calculated FT-IR and FT-Raman intensities are given in Table 9.4. Figure 9.2 and Figure 9.3 represent the experimental FT-IR and FT-Raman. The scaling of calculated Hartree–Fock (HF) frequencies were done with a multiplication factor of 0.989 below 1800 cm^{-1} and 0.9387 above 1800 cm^{-1} in order to match them with the experimental values.

9.4.3.1 Vibrations of SO_2

The SO_2 in-plane and out of plane vibrations for p-iodo benzene sulfonyl chloride are reported at 780, 694, 550 and 180 cm^{-1}, respectively [24, 25]. The SO_2 asymmetric stretching vibrations are at 1327 cm^{-1} in FT-IR for DPBS. The band observed at 1161 cm^{-1} in FT-IR and 1160 cm^{-1} in FT-Raman is assigned to SO_2 symmetric vibrations. They correlate well with the scaled HF values at 1335 cm^{-1} (mode no: 63) and 1158 cm^{-1} (mode no: 52), respectively, and with the literature values [25]. The SO_2 in-plane and out-of-plane vibrations are identified at 774 cm^{-1} and 527 cm^{-1} in FT-IR, respectively, for the title compound.

9.4.3.2 Vibrations of N–H

Heteroaromatic molecule containing N–H group shows its stretching absorption strongly and broadly in the region 3500–3220 cm^{-1} [26, 27]. For DPBS, the band at 3496 cm^{-1} in FT-IR is due to symmetric stretching vibration of N–H and it match with the computed HF scaled value at 3582 cm^{-1} (mode no: 93). The in plane bending vibrations of N–H bands usually appears in the range 1650–1580 cm^{-1} and N–H wagging appears in the range 900–650 cm^{-1}. The band in FT-IR at 1570 cm^{-1} denotes N–H in plane bending vibrations and the N–H out of plane vibration is noted at 681 cm^{-1} in FT-IR and at 676 cm^{-1} in FT-Raman. These vibrational modes correlate well with the scaled HF values.

9.4.3.3 Vibrations of C–Cl

The C–Cl stretching vibrations generally occur in the region 760–505 cm^{-1} [28]. For 1,3-dibromo-5-chlorobenzene, Arivazhagan et al. observed in-plane and out of plane vibrations of C–Cl at 352 cm^{-1} and 157 cm^{-1} in FT-Raman, respectively [29]. For the title compound DPBS, the band at 748 cm^{-1} in FT-IR and at 750 cm^{-1} in FT-Raman is attributed to C–Cl stretching vibrations. The in plane and out of plane vibrations of C–Cl is observed at 370 cm^{-1} and 254 cm^{-1} in FT-Raman, respectively.

9.4.3.4 Vibrations of C–S

The stretching modes of C–S are usually observed in the range 670–930 cm^{-1} [30, 31]. For DPBS, the C–S stretching mode is observed at 888 cm^{-1} in FT-IR and it matches with the computed value 885 cm^{-1} (mode no. 38) of HF/6-311++G(d,p) method. The in-plane vibration of C–S occurs at 580 cm^{-1} at the mode no. 26. The peak at 414 cm^{-1} in FT-Raman is due to C–S out of plane vibrations and it matches with scaled HF value 412 cm^{-1}.

9.4.3.5 Vibrations of C–C

The stretching modes of C–C in benzene derivatives are usually assigned between 1400 and 1650 cm^{-1} [31, 32]. In the present work, the vibrational frequencies observed in FT-IR at 1493, 1668 and 1714 cm^{-1} are observed due to the stretching vibrations of C–C. The bands at 1141 cm^{-1} and 688 cm^{-1} in FT-IR spectrum are noted for in plane and out of plane vibrations of C–C.

9.4.3.6 Vibrations of C–H

In aromatic compounds, the stretching vibrations of C–H occurs above 3000 cm^{-1} with a multiplicity of weak to moderate bands, compared with the C–C stretching of aliphatic compounds [33]. Roeges observed that the C–H stretching vibrations of the aromatic ring occur in the region 3120–3000 cm^{-1}. For DPBS, the band at 3332 cm^{-1} in FT-IR is assigned to C–C stretching vibrations. The computed values of these C–C stretching modes (mode nos: 85–92) for DPBS are found to be in the range from 3106–3180 cm^{-1} at scaled HF/6-311G(d,p) level of calculation. The in-plane and out-of-plane vibrations of C–H usually occur in the range of 1300–1000 cm^{-1} and 950–800 cm^{-1}, respectively [34–36]. In DPBS, the eight in-plane vibrations of C–H are identified at the range of 1216–1262 cm^{-1} and 1455–1538 cm^{-1} and the

Table 9.4: The calculated FT-IR and FT-Raman spectrum of DPBS.

Mode no.	Exptl		HF unscaled	HF	B3LYP	BP86	MO6	I_{IR}	I_{Raman}	Vibrational assignments
	IR	Raman								
1			15	14	15	15	7	0.0031	95.8576	Lattice vibrations
2			17	16	18	19	18	0.0035	97.6577	Lattice vibrations
3			27	26	25	26	27	0.5798	30.9683	Lattice vibrations
4			49	48	50	51	47	0.0233	3.3926	Lattice vibrations
5			62	61	57	56	57	0.0663	2.3191	Lattice vibrations
6			83	82	82	81	86	0.0610	1.2021	Lattice vibrations
7			118	117	110	107	117	0.0234	1.1785	Lattice vibrations
8			164	163	147	140	152	0.0031	0.0646	Lattice vibrations
9			188	186	169	162	169	1.3602	0.2392	γ SO_2
10			197	195	177	168	175	3.7126	0.3405	Lattice vibrations
11			205	203	188	183	197	1.5348	0.4581	Lattice vibrations
12			210	208	193	188	197	0.3483	0.2762	Lattice vibrations
13		254	259	257	233	224	238	2.8510	0.7420	γ C–Cl
14			267	265	240	228	239	3.2021	0.4698	Lattice vibrations
15			317	314	285	273	287	1.4775	0.2975	Lattice vibrations
16			349	346	317	304	323	0.0017	0.0624	Lattice vibrations
17		370	372	369	347	334	351	0.8405	0.0039	β C–Cl
18		414	415	412	367	350	372	0.0557	0.0217	γ C–S
19			434	431	400	387	404	1.2053	0.4262	γ C–Cl + γ C–C
20			450	446	412	394	406	2.2288	0.0004	γ C–H
21			473	469	423	404	429	8.1754	0.1988	γ C–H + γ N–H + γ SO_2
22	507		497	493	449	429	446	5.2655	0.0132	γ C–H + γ N–H
23	527		538	534	483	464	491	0.0037	0.0805	β SO_2
24			543	539	491	470	494	0.0508	0.1649	Lattice vibrations

25		552	553	549	502	480	507	0.0109	0.0603	Lattice vibrations
26	580		582	577	530	509	532	2.7941	0.1040	β C–S
27			613	608	561	542	568	3.0287	0.0951	γ C–C
28	638	648	652	647	582	556	585	0.0160	0.0101	γ C–C
29			657	652	612	587	615	44.7470	0.1356	γ C–C
30	681	676	677	672	635	614	626	0.9166	0.1577	γ N–H
31	688		697	691	637	631	646	18.1275	0.1852	γ C–C
32	748	750	741	735	683	660	683	0.0030	0.2964	$_{v}$C–Cl
33	774		770	764	709	675	704	0.0262	0.0043	$_{v}$C–Cl
34	817		792	786	711	686	705	9.6894	0.0111	ρCH_2
35	835	829	833	826	765	737	760	0.0191	0.1196	ρCH_2
36	856		845	838	784	757	781	68.1501	0.0164	γ C–H
37			875	868	809	783	815	5.4570	0.0915	γ C–H
38	888		892	885	819	794	827	0.1199	0.2918	$_{v}$C–S + γ C–H
39		941	943	935	841	799	834	99.8022	0.0071	γ C–H
40			948	940	855	821	848	8.0527	0.0202	γ C–H
41			971	963	866	832	873	0.9864	0.2661	γ C–H
42		997	995	987	883	835	884	12.5384	0.0005	γ C–H
43		1024	1025	1017	925	884	918	0.0002	0.0272	γ C–H
44	1055	1056	1081	1072	983	937	978	0.7987	0.7504	τCH_2
45			1098	1089	990	939	982	100.5181	0.0004	τCH_2
46		1100	1109	1100	1001	954	996	7.0030	0.0011	β C–C
47			1111	1102	1017	987	1009	0.2546	0.0057	β C–C
48			1113	1104	1037	999	1031	0.0077	0.0155	β C–C
49			1118	1109	1039	1004	1037	14.0934	0.2187	β C–C
50		1125	1131	1122	1047	1015	1046	3.9858	0.1305	β C–C
51	1141		1138	1129	1050	1021	1070	14.2837	0.4377	β C–C + β C–H
52	1161	1160	1167	1158	1070	1035	1092	88.9246	0.0074	$_{v}$ SO_2
53			1172	1163	1102	1060	1103	0.0048	0.0452	β N–H
54		1179	1184	1175	1113	1079	1104	247.9459	0.0566	β C–C + β C–H

(continued)

Table 9.4 (continued)

Mode no.	Exptl		HF unscaled	HF	B3LYP	BP86	MO6	I_{IR}	I_{Raman}	Vibrational assignments
	IR	Raman								
55			1216	1206	1116	1084	1132	21.8194	0.3371	β C–H
56			1219	1209	1132	1088	1142	7.3893	0.0753	β C–H
57			1224	1214	1164	1125	1156	1.5706	0.0342	β C–H
58	1244	1250	1262	1252	1182	1148	1178	0.1953	0.1627	β C–H
59	1277	1271	1276	1266	1203	1161	1180	11.8049	0.0271	ωCH_2
60			1288	1278	1205	1166	1195	107.5910	0.0276	ωCH_2
61		1291	1299	1289	1223	1188	1223	1.8751	0.0145	ωCH_2 + β C–H
62			1309	1299	1271	1222	1242	49.2345	0.1985	β C–H
63	1327		1346	1335	1301	1253	1294	2.4881	0.0082	$_{va}$ SO_2
64	1378	1375	1385	1374	1308	1270	1312	2.8732	0.0079	βC–H
65	1418	1410	1422	1411	1320	1276	1321	0.5401	0.0417	δCH_2
66			1440	1428	1323	1311	1335	1.9773	0.0503	δCH_2
67		1446	1455	1443	1343	1317	1344	0.0003	0.0479	β CCC + β C–H
68			1472	1460	1362	1337	1367	0.0332	0.0040	β CCC + β C–H
69	1493		1514	1502	1396	1339	1387	6.2366	0.0098	β CCC + β C–H
70			1535	1523	1398	1353	1394	5.8594	0.0807	β CCC + β C–H
71	1570		1599	1586	1450	1391	1434	0.0197	0.0027	β N–H
72		1599	1604	1591	1477	1422	1459	33.4310	0.0288	β CCC
73			1618	1605	1483	1436	1465	26.5595	0.0023	β CCC
74			1619	1606	1495	1440	1471	0.7699	0.1545	β CCC
75		1627	1641	1628	1516	1460	1477	5.4489	0.0287	β CCC
76	1668		1650	1637	1526	1477	1516	8.2003	0.0062	β CCC
77	1714		1733	1719	1590	1541	1605	63.9673	0.0497	β CCC
78			1763	1749	1605	1551	1617	0.7492	0.0893	β CCC,
79			1764	1750	1623	1575	1633	6.9333	0.3927	β CCC
80	1800		1790	1776	1644	1594	1656	0.5362	0.2674	β CCC + βC–H

81			3167	2973	3003	2923	2966	21.5799	0.0863	$_{v}$ CH_2
82	2981	2983	3177	2982	3026	2955	3020	3.0341	0.1607	$_{v}$ CH_2
83	3032	3034	3228	3030	3077	3006	3069	1.5017	0.0817	$_{va}$ CH_2
84	3066	3065	3272	3071	3109	3030	3085	39.7514	0.0206	$_{va}$ CH_2
85			3309	3106	3152	3075	3137	0.4227	0.0390	$_{va}$ C–H
86			3313	3110	3155	3078	3139	214.6154	0.1083	$_{va}$ C–H
87			3326	3122	3168	3092	3157	160.2061	0.1603	$_{va}$ C–H
88			3337	3132	3177	3101	3166	41.2531	0.0537	$_{va}$ C–H
89			3349	3144	3189	3113	3178	32.7635	0.4175	$_{v}$ C–H
90			3367	3161	3201	3119	3185	220.6578	0.0911	$_{va}$ C–H
91			3381	3174	3213	3135	3196	36.0844	0.0728	$_{v}$ C–H
92	3332		3388	3180	3215	3137	3199	9.6399	0.1443	$_{v}$ C–H
93	3496		3816	3582	3551	3427	3547	86.5178	0.0311	$_{v}$ N–H

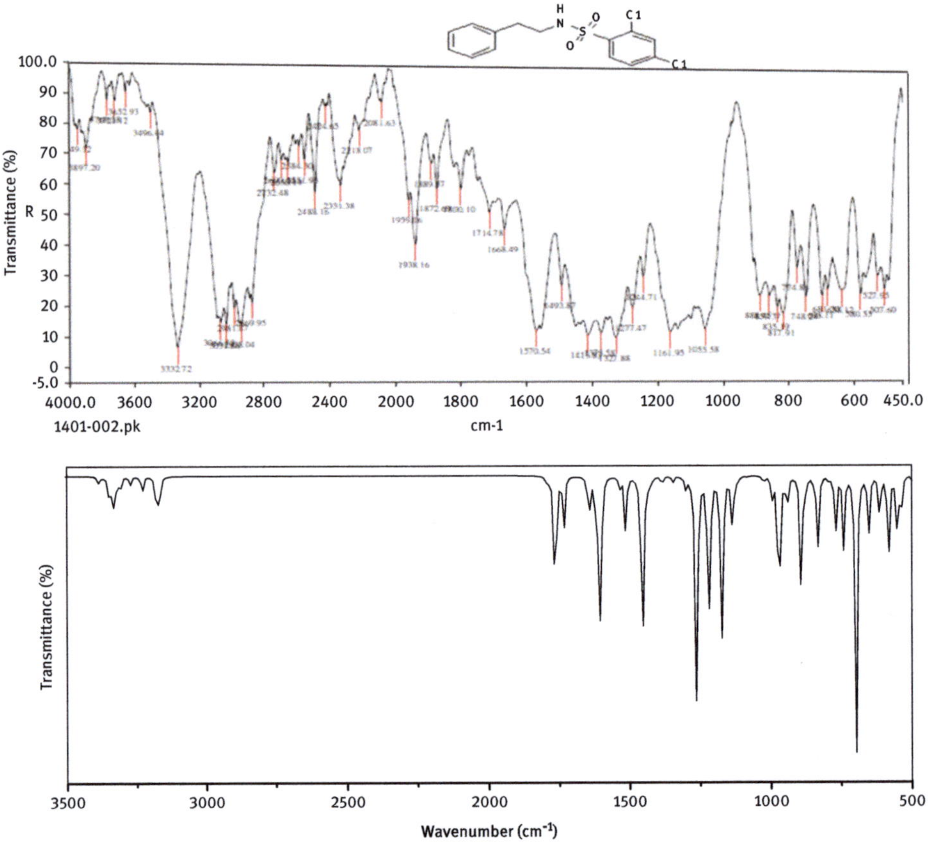

Figure 9.2: (a) Observed and (b) calculated FT-IR spectrum of DPBS molecule.

eight out-of-plane vibrations of C–H are noted at the range of 845–1025 cm^{-1} using scaled HF/6-311G(d,p) method. The band identified at 1244 and 1493 cm^{-1} in FT-IR and at 1250 cm^{-1} in FT-Raman is signified to in-plane vibrations of C–H. The peak at 856 cm^{-1} in FT-IR is assigned to out of plane vibrations of C–H.

9.4.3.7 Vibrations of methylene

In DPBS molecule, the asymmetric stretching vibrations of methylene are found at 3066 cm^{-1} in FT-IR and at 3065 cm^{-1} in FT-Raman. The peak identified at 2981 cm^{-1} in FT-IR and at 2983 cm^{-1} is due to methylene symmetric vibrations. These vibrations match with the computed HF values observed in the range of 2973–3071 cm^{-1} (mode nos: 81–84). The scissoring mode of the methylene group usually occurs in the region 1455–1380 cm^{-1} [36, 37]. The characteristic band at 1418 cm^{-1} in FT-IR, 1410 cm^{-1} in FT-Raman and the scaled HF frequency at 1411 cm^{-1} and 1428 cm^{-1} (mode nos: 65, 66) are corresponding to scissoring mode of methylene vibrations. The wagging vibration of

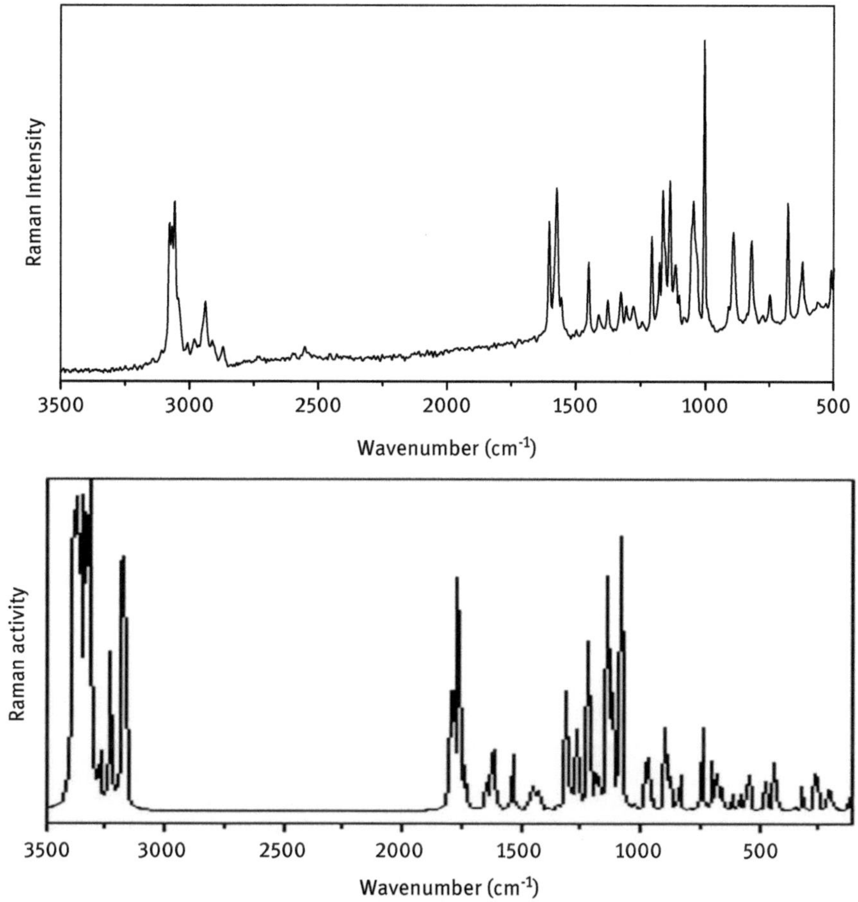

Figure 9.3: (a) Calculated and (b) observed FT-Raman spectrum of DPBS molecule.

CH_2 is noted at 1277 cm^{-1} in FT-IR and at 1271 cm^{-1} in FT-Raman. The band at 1055 cm^{-1} in FT-IR and at 1056 cm^{-1} in FT-Raman is due to twisting vibrations of methylene. These vibrational modes correlate with the computed values at mode nos: 59, 60 and 44, 45, respectively. The band at 817 cm^{-1} and 835 cm^{-1} in FT-IR were due to the rocking vibrations of methylene group. These results satisfy with the scaled frequencies (mode nos: 34, 35) and the literature data [38].

9.4.4 Molecular orbital (HOMO and LUMO)

The LUMO of the DPBS molecule is delocalized over the whole C–C bond of phenyl ring containing chlorine atoms and on the sulfonyl group. The HOMO is located above the NH group, methylene and on the benzene ring. The HOMO–LUMO transition indicates the transfer of electron density to the sulfonyl group and the phenyl ring containing chlorine atom. The molecular orbital picture for DPBS are shown in Figure 9.3. The HOMO–LUMO energy gap of DPBS was obtained using B3LYP/6-311++G(d,p) method as 5.21 eV. The energy gap indicates the charge transfer of the molecule through π conjugated system.

9.4.5 Global reactivity descriptors

The global reactivity descriptors such as electronegativity (χ), chemical potential (μ), global hardness (η), global softness (S) and electrophilicity index (ω) can be deduced from the density functional theory [39–41]. They are obtained from molecular orbitals ε_{HOMO} and ε_{LUMO} as follows:

$$\chi = -1/2(\varepsilon_{\mathrm{LUMO}} + \varepsilon_{\mathrm{HOMO}}) \tag{9.1}$$

$$\mu = -\chi = 1/2(\varepsilon_{\mathrm{LUMO}} + \varepsilon_{\mathrm{HOMO}}) \tag{9.2}$$

$$\eta = 1/2\,(\varepsilon_{\mathrm{LUMO}} - \varepsilon_{\mathrm{HOMO}}) \tag{9.3}$$

$$S = 1/2\eta \tag{9.4}$$

$$\omega = \mu^2/2\eta \tag{9.5}$$

In DPBS, the larger values of μ (Table 9.5) indicate that charge transfer processes are more predominant and this leads to stabilization through hyperconjugative interactions.

9.4.6 NLO effects

The first hyperpolarizability (β), dipole moment (μ) and polarizability (α) are calculated for DPBS molecule using the B3LYP/6-311++G(d,p) method from the following equations:

$$\alpha_0 = \frac{\alpha_{xx} + \alpha_{yy} + \alpha_{zz}}{3} \tag{9.6}$$

$$\alpha = 2^{-1/2}\left[(\alpha_{xx} - \alpha_{yy})^2 + (\alpha_{yy} - \alpha_{zz})^2 + (\alpha_{zz} - \alpha_{xx})^2 + 6\alpha_{xx}{}^2\right]^{1/2} \tag{9.7}$$

$$\beta_0 = \left(\beta_x^2 + \beta_y^2 + \beta_z^2\right)^{1/2} \tag{9.8}$$

$$\beta_x = \beta_{xxx} + \beta_{xyy} + \beta_{xzz} \tag{9.9}$$

$$\beta_y = \beta_{yyy} + \beta_{xxy} + \beta_{yzz} \tag{9.10}$$

Table 9.5: The electric dipole moment μ (D) the average polarizability α_{tot} ($\times10^{-24}$esu) and Thefirst hyperpolarizability β_{tot} ($\times10^{-31}$esu) of DPBS molecule.

Parameters	HF	B3LYP	BP86	M06
μ_x	1.724	1.543	1.478	1.547
μ_y	4.507	4.257	4.141	4.263
μ_z	1.454	1.364	1.321	1.388
μ	5.040	4.729	4.591	4.744
α_{xx}	5.949	5.886	5.073	4.294
α_{xy}	−13.776	−4.158	−4.341	−5.224
α_{yy}	7.826	−12.941	−12.373	−12.202
α_{xz}	−4.254	−0.515	−0.901	−0.338
α_{yz}	−0.385	−4.661	−4.552	−5.341
α_{zz}	−5.116	7.075	7.299	7.907
α_{tot} (esu)	23.145	21.932	20.622	19.999
β_{xxx}	−11.471	−12.736	−15.615	−26.253
β_{xxy}	30.415	26.809	28.833	31.685
β_{xyy}	36.540	31.834	29.876	30.023
β_{xyx}	37.262	39.163	36.999	42.107
β_{yyx}	−30.906	−29.933	−29.896	−24.775
β_{xyz}	5.512	4.433	4.446	2.471
β_{xzz}	19.997	18.700	18.653	19.795
β_{yyz}	−12.723	−14.635	−14.747	−11.353
β_{yyy}	10.442	18.700	8.568	9.370
β_{zzz}	9.415	9.340	8.979	10.676
β (esu)	4.3342E−31	4.2547E−31	4.0127E−31	4.6192E−31

$$\beta_z = \beta_{zzz} + \beta_{xxz} + \beta_{yyz} \tag{9.11}$$

and the total dipole moment

$$\mu = \left(\mu_x^2 + \mu_y^2 + \mu_z^2\right)^{1/2} \tag{9.12}$$

The parameters described above are listed in Table 9.6. The dipole moment (μ) was found to be 4.73 D. The calculated anisotropy of the polarizability is 21.932. The first hyperpolarizability value (β) is equal to 4.2547×10^{-31} esu. The dipole moment and first hyperpolarizability are nearly 3.4 and 1.2 times greater than urea. Hence the title compound can be further studied for exploring the non-linear optical properties.

9.4.7 Analysis of molecular electrostatic potential

Molecular electrostatic potential predicts the possibility of hydrogen bonding and also illustrates the structure-activity relationship of biomolecules and drugs [42–46]. The electrostatic potential surface of DPBS molecule is shown in Figure 9.5. The color

Table 9.6: Thermodynamic parameters.

Parameters	HF	B3LYP	BP86	MO6
Molecular mass	329.004	329.004	329.004	329.004
Rotational constants (GHz)				
A	0.348	0.337	0.326	0.317
B	0.132	0.132	0.135	0.148
C	0.114	0.113	0.114	0.123
ZPVE (J/mol)	691740.7	644672.3	624635.6	643071.8
ZPVE (kcal/mol)	165.3	154.1	149.3	153.7
Thermal energy	−2058.260	−2065.044	−2065.149	−2064.423
Enthalpy	−2058.327	−2065.048	−2065.148	−2064.422
Free energy	−2058.327	−2065.118	−2065.220	−2064.493
SCF energy (a.u)	−2058.541	−2065.313	−2065.406	−2064.687
Total energy (kcal/mol)	176.239	165.770	161.331	165.349
Cv (calmol^{-1}K)	64.727	69.850	72.153	69.802
S (calmol^{-1}K)	143.441	148.201	150.254	149.073
HOMO	−213.780	−164.357	−148.280	−170.745
LUMO	20.249	−44.289	−61.797	−40.348
Electronegativity (χ)	96.765	104.323	105.038	105.547
Chemical potential (μ_p)	−96.765	−104.323	−105.038	−105.547
Global hardness (η)	117.014	60.037	43.241	65.198
Global softness (S)	1682.542	3279.492	4553.087	3019.891
Electrophilicity index (ω)	40.010	90.6124	127.5727	85.4041

scheme for the MESP surface is red, electron rich, partially negative charge; blue, electron deficient, partially positive charge; light blue, slightly electron deficient region; yellow, slightly electron rich region; green, neutral, respectively. From the Figure 9.4, it can be understood that the electrophilic region is over the aromatic ring, methylene and NH group and the nucleophilic regions are over over the phenyl ring containing chlorine atom.

9.4.8 NBO analysis

NBO analyses is a useful tool for understanding delocalization of electron from donor NBOs to acceptor NBOs within the molecule [47–51]. The bond pair-bond pair interactions and lone pair-bond pair interactions with stabilization energy greater than 13 kJ/mol are listed in Table 9.7 and Table 9.8.

The orbital overlap between *n*(Cl) and σ* (C–C) bond leads to intramolecular charge transfer and stabilization of the system. The n–π conjugation between the O and Cl lone pair electrons and benzene ring π system is evident in the ground state.

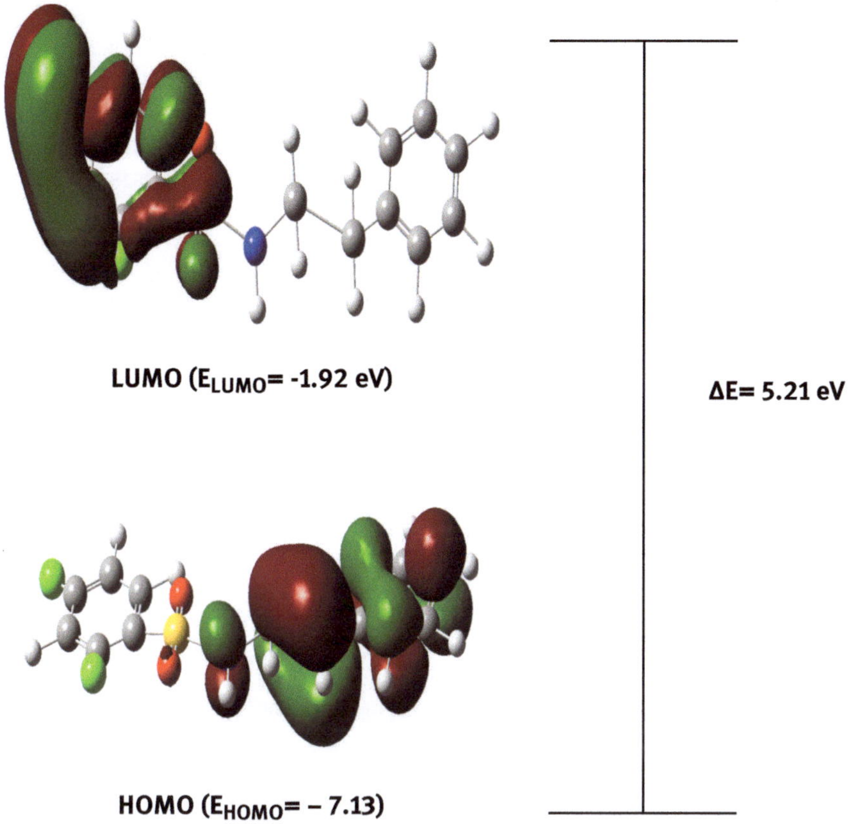

Figure 9.4: The calculated frontiers energies of DPBS molecule.

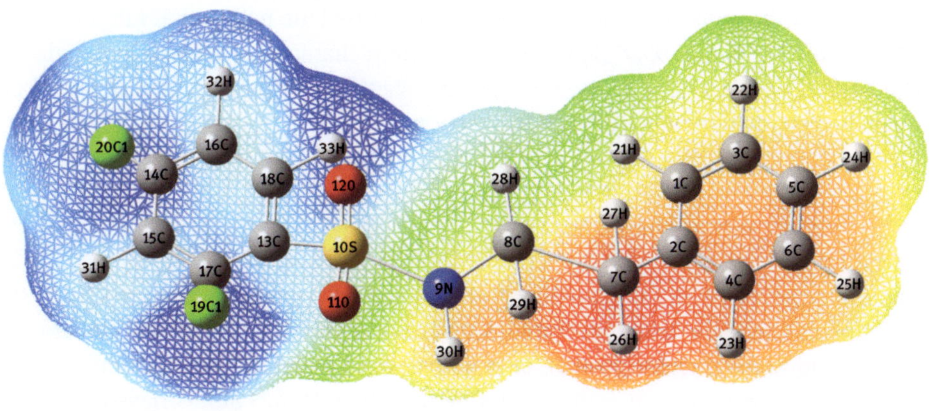

Figure 9.5: The calculated MEP map of DPBS molecule.

Table 9.7: The second order perturbation energies *E*(2) (kcal/mol) corresponding to the most important charge transfer interactions (donor–acceptor) of DPBS using B3LYP functional.

Donor NBO (i)	ED (i) (e)	Acceptor NBO (j)	ED (j) (e)	E (2)	E(j)–E(i) a.u	F(i,j) a.u
σC4–C5	1.656	σ*C7–C9	0.330	20.98	0.28	0.069
		σ*C11–C13	0.327	19.65	0.28	0.067
σC7–C9	1.663	σ*C4–C5	0.348	19.87	0.28	0.067
		σ*C11–C13	0.327	20.62	0.28	0.068
σC11–C13	1.667	σ*C4–C5	0.348	21.46	0.29	0.070
		σ*C7–C9	0.330	20.02	0.28	0.067
σC21–C28	1.968	σ*C22–C23	0.373	22.51	0.28	0.071
		σ*C25–C26	0.384	17.31	0.28	0.063
σC22–C23	1.678	σ*C21–C28	0.371	17.16	0.29	0.064
		σ*C25–C26	0.384	19.33	0.29	0.068
σC25–C26	1.657	σ*C21–C28	0.371	22.38	0.29	0.072
		σ*C22–C23	0.373	20.61	0.28	0.068
LP(3) Cl2	1.914	σ*C22–C23	0.373	13.03	0.32	0.063
LP(3) Cl3	1.913	σ*C25–C26	0.384	13.38	0.32	0.064
LP(2) O30	1.805	σ*S1–C21	0.220	17.67	0.42	0.077
		σ*S1–N32	0.227	11.53	0.43	0.064
LP(3) O30	1.786	σ*S1–O31	0.141	20.38	0.57	0.098
LP(2) O31	1.817	σ*S1–C21	0.220	17.82	0.42	0.078
		σ*S1–N32	0.227	13.23	0.43	0.068
σC22–C23	1.678	σ*C21–C28	0.371	282.60	0.01	0.078

The LP(3)Cl2→C22–C23, LP(3)Cl3→C25–C26, LP(3)O30→S1-O31, LP(2)O31→S1–C21 energies are 13.03, 13.38, 20.38 and 17.82 kJ/mol, respectively. It is evident from these values that n–π conjugation takes between Cl, O and benzene ring. In DPBS compound, the interaction among C22–C23→C21–C28 results in a stabilization energy of 282.6 kJ/mol. This strong stabilization leads to the larger delocalization and induces the large bioactivity of the molecule. The p-character of chlorine lone pair orbital LP(3) Cl2 and LP(3) Cl3 are 99.95 and 99.97%, respectively. The p-character of oxygen lone pair orbital LP(2) O30 and LP(2) O31 are 99.92%. This signifies the participation of lone pair orbital in electron donation in the DPBS compound.

9.4.9 Mulliken atomic charge analysis

Atomic charges of DPBS calculated by Mulliken and NBO methods using the HF/6-311++G(d,p) method are listed in Table 9.9. The magnitudes of the carbon atomic charges, found to be either positive or negative, were noted to change from −0.002 e to 1.106 e. All the hydrogen, sulfur, chlorine and C4, C22, C25 atoms have a positive charge. The sulfur (S1) atom has the maximum positive charge (0.526 e in HF method) than the other carbon atoms; and they represent as an acceptor atom. It is also

Table 9.8: NBO results showing the formation of Lewis and non-Lewis orbitals of DPBS.

Bond (A–B)	ED/energy(a.u)	ED_A %	ED_B %	NBO	s %	p %
σ	1.656	49.54	50.46	0.703 ($sp^{99.99}$)	0.02	99.93
C4–C5	−0.258			0.710 ($sp^{1.00}$)	0.00	99.95
σ	1.662	49.69	50.31	0.704 ($sp^{1.00}$)	0.00	99.96
C7–C9	−0.258			0.709 ($sp^{1.00}$)	0.00	99.96
σ	1.665	49.62	50.38	0.704 ($sp^{1.00}$)	0.00	99.96
C11–C13	−0.258			0.709 ($sp^{1.00}$)	0.00	99.96
σ	1.677	56.31	43.69	0.750 ($sp^{1.00}$)	0.00	99.97
C21–C28	−0.302			0.661 ($sp^{1.00}$)	0.00	99.94
σ	1.678	51.46	48.54	0.717 ($sp^{1.00}$)	0.00	99.98
C22–C23	−0.309			0.696 ($sp^{1.00}$)	0.00	99.95
σ	1.658	51.60	48.40	0.718 ($sp^{1.00}$)	0.00	99.97
C25–C26	−0.304			0.695 ($sp^{1.00}$)	0.00	99.94
LP(3) Cl2	1.919			$sp^{1.00}$	0.01	99.95
	−0.349					
LP(3) Cl3	1.918			$sp^{1.00}$	0.01	99.97
	−0.343					
LP(2) O30	1.807			$sp^{1.00}$	0.01	99.92
	−0.304					
LP(3) O30	1.780			$sp^{1.00}$	0.01	99.87
	−0.303					
LP(2) O31	1.817			$sp^{1.00}$	0.01	99.92
	−0.309					
LP(3) O31	1.786			$sp^{99.99}$	0.06	99.85
	−0.307					
π	1.678	51.46	48.54	0.717 ($sp^{1.00}$)	0.00	99.98
C22–C23	−0.309			0.696 ($sp^{1.00}$)	0.00	99.95

observed that C21 and C26 atoms have maximum negative charge and hence they represent as donor atoms. The charge values for S1, Cl2 and Cl3 are found to be 0.204, 0.402, 0.509 using the B3LYP/6-311++G(d,p) method. The hydrogen atomic charges are found in an order from 0.154 (H19) to 0.229 (H24), 0.131 (H19) to 0.250 (H24), 0.136 (H19) to 0.258 (H24) and −0.008 (H19) to 0.276 (H24) at the HF level of theory and the B3LYP, BP86 and M06 functionals.

9.5 Summary

The molecular structure and vibrational spectroscopic analysis of DPBS have been studied by FT-IR and FT-Raman spectroscopy. A complete vibrational and molecular analyses of DPBS have been obtained using theoretical methods. The calculated geometrical parameters and vibrational frequencies obtained from density functional theory calculations are in good agreement with the experimental values obtained for

Table 9.9: Atomic charges of DPBS.

Atoms	HF		B3LYP		BP86		MO6	
	Mulliken	NBO	Mulliken	NBO	Mulliken	NBO	Mulliken	NBO
S1	0.525	2.481	0.203	2.226	−0.016	2.129	−0.098	2.270
Cl2	0.298	0.013	0.401	0.032	0.472	0.043	0.474	0.034
Cl3	0.460	0.007	0.508	0.029	0.585	0.014	0.572	0.032
C4	1.106	−0.034	0.872	−0.036	0.912	−0.038	0.577	−0.031
C5	−0.323	−0.193	−0.330	−0.201	−0.348	−0.208	−0.287	−0.202
H6	0.181	0.187	0.143	0.201	0.152	0.208	0.159	0.198
C7	−0.349	−0.176	−0.297	−0.195	−0.330	−0.203	−0.293	−0.191
C9	−0.508	−0.195	−0.373	−0.205	−0.419	−0.211	−0.437	−0.203
C11	−0.264	−0.175	−0.227	−0.193	−0.259	−0.202	−0.173	−0.189
C13	−0.215	−0.190	−0.241	−0.199	−0.277	−0.205	−0.088	−0.201
C15	−0.423	−0.343	−0.148	−0.398	−0.087	−0.421	−0.232	−0.400
C18	−0.916	−0.101	−0.917	−0.179	−1.006	−0.210	−0.553	−0.180
C21	−1.295	−0.352	−0.778	−0.305	−0.589	−0.301	−0.723	−0.323
C22	0.383	0.013	0.137	−0.024	0.0327	−0.038	0.229	−0.017
C23	−0.391	−0.238	−0.387	−0.239	−0.395	−0.247	−0.555	−0.242
C25	−0.001	0.017	0.037	−0.013	0.002	−0.026	−0.001	−0.006
C26	−0.452	−0.220	−0.512	−0.219	−0.579	−0.226	−0.532	−0.220
C28	0.033	−0.117	−0.107	−0.160	−0.176	−0.175	−0.118	−0.155
O30	−0.112	−1.001	−0.017	−0.892	0.042	−0.845	0.025	−0.904
O31	−0.283	−1.021	−0.203	−0.917	−0.154	−0.873	−0.147	−0.924
N32	−0.194	−0.926	−0.076	−0.865	−0.001	−0.844	−0.127	−0.887
H19	0.143	0.153	0.130	0.179	0.136	0.191	−0.007	0.179
H24	0.309	0.228	0.250	0.240	0.258	0.248	0.275	0.237
H33	0.293	0.400	0.246	0.398	0.243	0.399	0.249	0.404

the DPBS molecule. HOMO–LUMO energy gap signifies the charge transfer within the compound. The dipole moment, polarizability and first order polarizabilities suggest that DPBS can be explored for its NLO properties. NBO analysis indicates the molecular interaction taking place within the molecule. MEP study identifies the various electrophilic and nuclophilic regions of DPBS compound.

Acknowledgements: The authors acknowledge the facilities from their respective Universities.

References

[1] Jagrut VB, Waghmare RA, Mane RA, Jadhav WN. An improved synthetic route for the synthesis of sulfonamides. Int J Chem Tech Res. 2011;3:1592.

[2] Supuran CT. Carbonic Anhydrase Inhibitors-Part 57:Quantum Chemical QSAR of a Group of 1,3,4-thiadiazole and 1,3,4-thiadiazoline Disulfonamides with Carbonic Anhydrase. Eur J Med Chem. 1999;34:41.

[3] Selvaag Edgar, Anholt Helle, Moan Johan, Thune Per. Inhibiting effects of antioxidants on drug-induced phototoxicity in cell cultures Investigations with sulphonamide-derived oral antidiabetics and diuretics. Journal of Photochemistry and Photobiology B: BiologyJournal. 1997;38:88–93.

[4] Supuran CT, Scozzaf A. Carbonic Anhydrase:its Inhibitors and Activators. Eur J Med Chem. 1999;34:799.

[5] Patel NB, Patel VN, Patel HR, Shaikh FM, Patel JC. Synthesis and Microbial Studies of (4-Oxo-thiazolidinyl) Sulphonamides Bearing Quinazolin-4-(3*H*)ones. Acta Poloniae PharmaceuticaActa. 2010;67(3):267–275.

[6] Supuran CT. Carbonic Anhydrase Inhibitors-Part 53-Synthesis of Substituted-Pyridinium Derivatives of Aromatic Sulfonamides:The First Non-Polymeric Membrane-Impermeable Inhibitors with Selectivity for Isozyme IV. Eur J Med Chem. 1998;31:577.

[7] Boyd AE. Sulfonylurea Receptors, Ion Channels and Fruit Flies, Diabetes. Diabetes. 1988;37:847.

[8] Supuran CT. Recent Developments of Carbonic Anhydrase Inhibitors as Potential Anticancer Drugs. J Med Chem. 2008;51:3051–3056.

[9] Ozbek N, Katircioglu H, Karacan N, Baykal T. N-Ethyl-N-(2-methoxyphenyl)benzene sulphonamide. Bioorg Med Chem. 2007;15:5105–9.

[10] Ghorab MM, Ragab FA, Hamed MM. Design, Synthesis and Anticancer Evaluation of Novel Tetrahydroquinoline Derivatives Containing Sulfonamide Moiety. Eur J Med Chem. 2009;44:4211–17.

[11] Borne RF, Peden RL, Waters IW, Weiner M, Jordan R, Coats EA. Anti-Inflammatory Activity of para-Substituted N-benzenesulfonyl Derivatives of Anthranilic Acid. J Pharm Sci. 1974;63:615–17.

[12] Bhat MA, Imran M, Khan SA, Siddiqui N. Biological Activities of Sulfonamides. Indian J Pharm Sci. 2005;67:151–9.

[13] Babu CS, Kavitha HP. 2,4-Dichloro-N-phenethyl benzenesulphonamide. Act Cryst. 2009;E65: o921.

[14] Frisch MJ, Trucks GW, Schlegel HB, Scuseria GE, Robb MA, Cheeseman JR, et al. GAUSSIAN 09, Revision D.01. Wallingford CT: Gaussian, Inc., 2013.

[15] Becke AD. Density-Functional Thermochemistry,III. The Role of Exact Exchange. J Chem Phys. 1993;91:5648.

[16] Lee C, Yang W, Parr RG. Development of the Colle-Salvetti Correlation-Energy Formula into a Functional of the Electron Density. Phys Rev B. 1998;37:785.

[17] Raghavachari K, Trucks GW. Highly Correlated Systems:Excitation Energies of First Row Transition Metals Sc-Cu. J Chem Phys. 1989;91:1062.

[18] Wong MW, Wilberg KB, Frisch MJ. Solvent effects. Tautomeric Equilibria of Formamide and 2-pyridone in the Gas Phase and Solution: an ab initio SCRF study. J Am Chem Soc. 1992;114:1645.

[19] Zhao Y, Truhlar DG. The MO6 Suite of Density Functionals for Main Group ThermoChemistry, Thermochemical Kinetics, Noncovalent Interactions, Excited States and Transition Elements: Two New Functionals and Systematic Testing of Four MO6-Class Functionals and 12 other Functionals. Theor Chem Accounts. 2008;120:215.

[20] Pulay P, Fogarasi G, Pongar G, Boggs JE, Varsha A. Combination of Theoritical ab initio and Experimental Information to Obtain Reliable Harmonic Force Constants. Scaled Quantum Mechanical (QM) Force Fields for Glyoxal, Acrolein, Butadiene, Formaldehyde and Ethylene. J Am Chem Soc. 1983;105:7073.

[21] Fogarasi G, Zhou X, Jaylor PW, Pulay P. The Calculation of ab initio Molecular Geometries: Efficient Optimization by Natural Internal Coordinates and Empirical Correction by Offset Forces. J Am Chem Soc. 1992;114:819.

[22] Silverstein M, Bassler GC, Morill C. Spectrocopic identification of organic compounds. New York: John Wiley, 1981.
[23] Wilson EB, Decius JC, Cross PC. Molecular vibrations. New York: Dover Publ. Inc., 1980.
[24] Roeges NP. A guide to the complete interpretation of infrared spectra of organic structures. New York: Wiley, 1994.
[25] Arivazhagan M, Prabhakaran S, Gayathri R. Molecular Structure, VIbrational Spectroscopic, First HyperPolarizability, NBO and HOMO of P-Iodobenzene Sulfonyl Chloride. Spectrochim ActaA. 2011;82:332–9.
[26] Erdogdu Y, Gulluoglu MT, Yurdakul S. Molecular Structure and Vibrational Spectra of 1,3-bis (4-piperidyl)Propane by Quantum Chemical Calculations. J Mol Struct. 2008;889:361–70.
[27] Erdogdu Y, Gulluoglu MT. Analysis of Vibrational Spectra of 2 and 3-Methylpiperidine Based on Density Functional Theory Calculations. Spectrochim Acta. 2009;74A:361–70.
[28] George S. Infrared and characteristic group frequencies tables and charts, 3rd ed. Chichester: Wiley, 2001.
[29] Arivazhagan RM. Quantum Chemical Studies on Structure of 1,3-dibromo-5-Chlorobenzene. Spectrochim ActaA. 2011;82:316–26.
[30] Colthup NB, Daly LH, Wiberly SE. Introduction to infrared and Raman spectroscopy, 3rd ed. Boston, MA: Academic Press, 1990.
[31] Socrates G. Infrared characteristic group frequencies. New York: John Wiley and Sons, 1980.
[32] Sathyanarayana DN. Vibrational spectroscopy theory and applications. New Delhi: New Age International Publishers, 2004.
[33] Coates J. Interpretation of infrared spectra: A practical approach. New York: John Wiley & Sons Inc, 2000.
[34] Krishnakumar V, Prabhavathi N. Spectroscopic (FT-IR, FT-Raman, UV and NMR) Investigation, Conformational Stability, NLO Properties, HOMO-LUMO and NBO Analysis of Hydroxyquinoline Derivatives by Density Functional Theory Calculations. Spectrochim Acta A. 2008;71:449.
[35] Singh SJ, Pandey SM. Vibrational Spectra of 2-Amino-6-nitrotoluene. Indian J Pure Appl Phys. 1974;12:300.
[36] Altun A, Golcuk K, Kumru M. Structure and Vibrational Spectra of p-methylaniline: Hartree-Fock, MP2 and Density Functional Theory Studies. J Mol Struct (Theochem). 2003;637:155.
[37] Lin-Vein D, Colthup NB, Fateley WG, Grasseli JG. The hand book of infrared and Raman characteristic frequencies of organic molecules. San Diego: Academic Press, 1991.
[38] Subhashchandrabose S, Saleem H, Ergodu Y, Rajarajan G, Thanikachalam V. FT-Raman, FT-IR spectra and Total Energy Distribution of 3-pentyl-2,6-diphenylpiperidin-4-one:DFT method. Spectrochim ActaA. 2011;82:260–9.
[39] Parthasarathi R, Padmanabhan J, Subramanian V, Maiti B, Chattraj PK. Chemical Reactivity Profiles of Two-Selected Poly Chlorinated Biphenyls. J Phys Chem A. 2008;107:10346–52.
[40] Parr RG, Szentpaly LV, Liu SJ. Electrophilicity Index. J Am Chem Soc. 1999;121:1922.
[41] Singh RN, Kumar A, Tiwari RK, Rawat P. A Combined Experimental and Theoritical (DFT and AIM) Studies on Synthesis, Molecular Structure, Spectroscopic Properties and Multiple Interactions Analysis in a Novel Ethyl-4-[2-(thiocarbamoyl)hydrazinylidene]-3,5,dimethyl-1H-pyrrole-2-carboxylate and its Dimer. Spectrochim Acta. 2013;112:182–90.
[42] Murray JS, Sen K. Molecular electrostatic potentials, concepts and applications. Amsterdam: Elsevier, 1996.
[43] Fleming I. Frontier orbitals and organic chemical reactions. New York: John Wiley and Sons, 1976:5–27.

[44] Seminario JM. Recent developments and applications of modern density functional theory Vol. 4. Elsevier, 1996:800–6. Amsterdam.
[45] Yesikaynak T, Binzet G, Mehmet Emen F, Florke U, Kulcu N, Arsian H. Theoretical and Experimental Studies on N-(6-methylpyridin-2-yl-carbamothioyl)biphenyl-4-Carboxamide. Eur J Chem. 2010;1:1–5.
[46] Chidangil S, Shukla MK, Mishra PC. A Molecular Electrostatic Potential Mapping Study of some Fluoroquinolone Anti-bacterial Agents. J Mol Model. 1998;4:250–8.
[47] Reed AE, Weinhold F. Natural Localized Molecular Orbitals. J Chem Phys. 1985;83:1736–40.
[48] Reed AE, Weinstock RB, Weinhold F. Natural Population Analysis. J Chem Phys. 1985;83:735–46.
[49] Reed AE, Weinhold F. Natural Bond Orbital Analysis of Near-Hartree-Fock Water Dimer. J Chem Phys. 1983;78:4066–73.
[50] Arjunan V, Raj A, Santhanam R, Marchewka MK, Mohan S. Structural, vibrational, electronic investigations and quantum chemical studies of 2-amino-4-methoxybenzothiazole. Spectrochim Acta A. 2013;102:327–40.
[51] Kavipriya R, Kavitha HP, Karthikeyan B, Natraj A. Molcular Structure, Spectroscopic (FT-IR, FT-Raman), NBO Analysis of N,N'-Diphenyl-6-piperidiny-1-yl-[1,3,5]-triazine-2,4-diamine. Spectrochim Acta A. 2015;156:476–87.

Index

https://doi.org/10.1515/9783110631623-010

Bei Fragen zur Produktsicherheit wenden Sie sich bitte an:
If you have any questions regarding product safety,
please contact:

Walter de Gruyter GmbH
Genthiner Straße 13
10785 Berlin
productsafety@degruyterbrill.com